ROBOTIC OBSERVATORIES

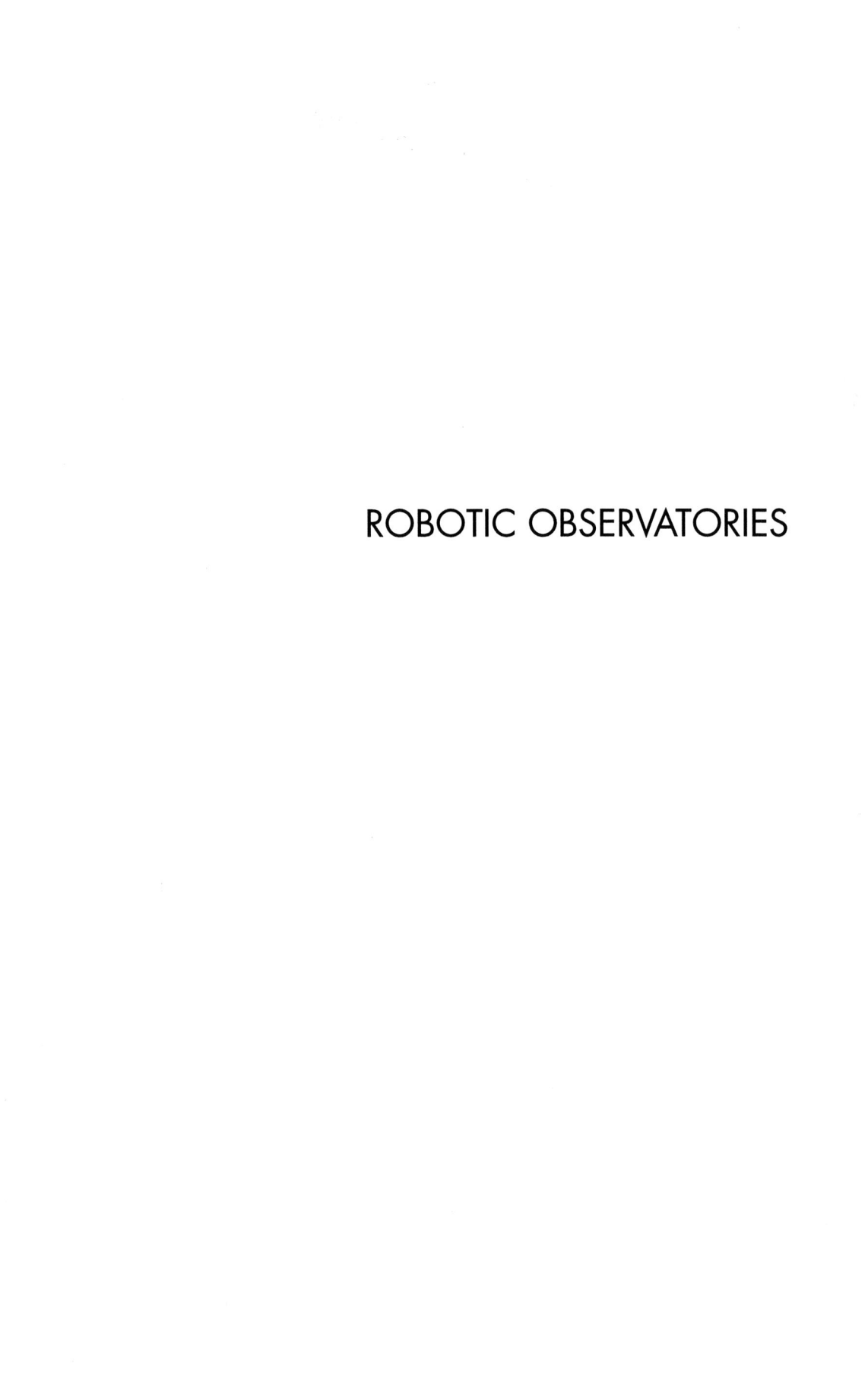

WILEY-PRAXIS SERIES IN ASTRONOMY AND ASTROPHYSICS
Series Editor: JOHN MASON, B.Sc., Ph.D.

This series aims to coordinate a team of international authors of the highest reputation, integrity and expertise in all aspects of astronomy and astrophysics. It makes a valuable contribution to the existing literature, encompassing all areas of both observational and theoretical astronomical research. The titles are illustrated with both black and white and colour photographs, and include many line drawings and diagrams, with tabular data, appendices and extensive bibliographies.

Aimed at a wide readership, the books will appeal to professional astronomers, astrophysicists and cosmologists, physicists and space scientists, postgraduate and undergraduate students in physics, astrophysics and astronomy and, in certain instances, amateur astronomers, high-flying 'A' level students, and non-scientists with a keen interest in astronomy.

ROBOTIC OBSERVATORIES
Michael F. Bode, Professor of Astrophysics and Assistant Provost for Research, Liverpool John Moores University, UK

THE AURORA: Sun-Earth Interactions
Neil Bone, School of Biological Sciences, University of Sussex, Brighton, UK

PLANETARY VOLCANISM: A Study of Volcanic Activity in the Solar System
Peter Cattermole, formerly Lecturer in Geology, Department of Geology, Sheffield University, UK, now Principal Investigator with NASA's Planetary Geology and Geophysics Programme

DIVIDING THE CIRCLE: The Development of Critical Angular Measurement in Astronomy 1500-1850
Allan Chapman, Wadham College, University of Oxford, UK

THE DUSTY UNIVERSE
Aneurin Evans, Department of Physics, University of Keele, UK

MARS AND THE DEVELOPMENT OF LIFE
Anders Hansson, Ph.D.

ASTEROIDS: Their Nature and Utilization
Charles T. Kowal, Computer Sciences Corp., Space Telescope Science Institute, Baltimore, Maryland, USA

COMET HALLEY - Investigations, Results, Interpretations
Volume 1: Organization, Plasma, Gas
Volume 2: Dust, Nucleus, Evolution
Editor: John Mason, B.Sc., Ph.D.

ELECTRONIC AND COMPUTER-AIDED ASTRONOMY: From Eyes to Electronic Sensors
Ian. S. McLean, Department of Astronomy, University of California at Los Angeles, California, USA

URANUS: The Planet, Rings and Satellites
Ellis D. Miner, Cassini Science Manager, NASA Jet Propulsion Laboratory, Pasadena, California, USA

THE PLANET NEPTUNE: An Historical Survey, Before Voyager
Patrick Moore, CBE, D.Sc.(Hon.)

ACTIVE GALACTIC NUCLEI
Ian Robson, Director, James Clerk Maxwell Telescope, Head Joint Astronomy Centre, Hawaii, USA

ASTRONOMICAL OBSERVATIONS OF ANCIENT EAST ASIA
Richard Stephenson, Department of Physics, University of Durham, UK, Zhentao Xu, Purple Mountain Observatory, Academia Sinica, Nanjing, China, Yaotiao Tiang, Department of Astronomy, Nanjing University, China

EXPLORATION OF TERRESTRIAL PLANETS FROM SPACECRAFT: Instrumentation, Investigation, Interpretation
Yuri A. Surkov, Chief of the Laboratory of Geochemistry of Planets, Vernadsky Institute of Geochemistry, Russian Academy of Sciences, Moscow, Russia

THE HIDDEN UNIVERSE
Roger J. Tayler, Astronomy Centre, University of Sussex, Brighton, UK

ROBOTIC OBSERVATORIES

Editor:
Michael F. Bode
Professor of Astrophysics and Assistant Provost for Research,
Liverpool John Moores University

JOHN WILEY & SONS
Chichester • New York • Brisbane • Toronto • Singapore

Published in association with
PRAXIS PUBLISHING
Chichester

Copyright © 1995 Praxis Publishing Ltd
The White House,
Eastergate, Chichester,
West Sussex, PO20 6UR, England

Published in 1995 by
John Wiley & Sons Ltd in
association with Praxis Publishing Ltd

All rights reserved.

No part of this book may be reproduced by any means,
or transmitted, or translated into a machine language
without the written permission of the publisher.

Wiley Editorial Offices

John Wiley & Sons Ltd, Baffins Lane,
Chichester, West Sussex PO19 1UD, England

John Wiley & Sons, Inc., 605 Third Avenue,
New York, NY 10158-0012, USA

Jacaranda Wiley Ltd, G.P.O. Box 859, Brisbane
Queensland 4001, Australia

John Wiley & Sons (Canada) Ltd, 22 Worcester Road,
Rexdale, Ontario M9W 1L1, Canada

John Wiley & Sons (SEA) Pte Ltd, 37 Jalan Pemimpin 05-04,
Block B, Union Industrial Building, Singapore 2057

A catalogue record for this book is available from the British Library

ISBN 0-471-95690-2

Printed and bound in Great Britain by Hartnolls Ltd, Bodmin

CONTENTS

Editor's introduction — vii

PART I ROBOTIC OBSERVATORIES IN OPERATION, UNDER DEVELOPMENT AND PLANNED

A review of European developments in automated and robotic telescopes for photometry — 3
J.E.F. Baruch

A progress report on the Sutherland automatic telescope — 11
D. Kilkenny

Developments in automatising a Celestron-based APT — 15
R. Hudson, G. Hudson and E. Budding'

The development of a robotic telescope system at Bradford — 21
J.E.F. Baruch

A project for an automated telescope in Argentina — 27
J.R. Garcia, S.A. Dominguez and J. Campos

The Fastnet Observatory CCD APT system — 35
E. Ansbro and H. van Bellingen

PART II OBSERVATIONAL RESULTS

The quest for precision robotic photometry — 41
G.W. Henry and D.S. Hall

The law of starspot lifetimes — 49
D.S. Hall and G.W. Henry

On-going studies of R CrB and UU Her stars with robotic telescopes — 59
J. D. Fernie

Corner stars for calibration of the Sky Survey fields—results from a pilot programme — 63
P.W. Hill, P. O'Neill and B.M. Lasker

The automatic 60-cm telescope of the Belogradchik Observatory—first results — 69
A. Antov and R. Konstantinova-Antova

PART III ROBOTIC TELESCOPE NETWORKS

Some GNAT issues   77
    D.L. Crawford

A complementary network to GNAT: An Arabian and French project for 3T1M automated photometric stations   85
    F.R. Querci, M. Querci, S. Kadiri and L. de Rancourt

Development of an academic network for astronomical use in Bulgaria   89
    P. Delchev and M. Tsvetkov

Facilities for the co-ordination of multi-site and multi-wavelength observing campaigns   95
    C.J. Butler and S.J. Magorrian

PART IV TECHNOLOGY AND IDEAS

Monitoring of active galactic nuclei   101
    I.G. van Breda

The concept of an APT network as a driver for a metrological reform in astronomical photometry   107
    C. Sterken

How accurate are photometric standards?   113
    T. Oja

Towards robotic IR observatories: improved IR passbands   117
    E.F. Milone, C.R. Stagg and A.T. Young

Liquid mirror telescopes   125
    E.F. Borra, R. Content and L. Girard

Parallelism in telescope and instrument control systems   131
    I.G. van Breda

Editor's summary   139
    M.F. Bode

Appendix: List of Participants in the Kilkenny Workshop   147

Subject index   151

Editor's introduction

M.F. Bode

*Astrophysics Group, Liverpool John Moores University,
Byrom Street, Liverpool, L3 3AF, UK.*

1. Why robotic observatories?

The progress of astronomy is inextricably linked to technological advance. Perhaps the most obvious example of this was the first use of the telescope at the beginning of the 17th century. In more recent times, the opening up of the electromagnetic spectrum to astronomical investigation has fundamentally changed our view of the cosmos. We are now entering an era in which the way we conduct ground-based observations is being revolutionised by a marrying of computer power, electronic communication, and detector technology. The end product is the fully robotic observatory, and in the pages of this book the reader will find details from around the world of the current state-of-the-art in this growing field, together with plans for its future development.

The push for the development of automated and robotic observatories has come from both ends of the spectrum of practising astronomers. The amateur community has become increasingly sophisticated in terms of the observational programmes that its members have embarked upon, and indeed they remain a valuable source of data to their professional counterparts. Amateurs have also gained access to more sophisticated hardware, particularly in respect of computers and detectors (e.g., CCD cameras) as prices for these have tumbled. With an automated telescope the amateur can set observing programmes in train, sleep and collect the results the next day and therefore still be fit to undertake their primary daytime employment.

At the other end of the spectrum is the professional astronomer who can always devour far more telescope time than is available on the world's large telescopes. Through their flexibility and efficiency, remotely operated robotic telescopes in fact open up many possibilities for the professional in the quest to understand more about the workings of our universe. As emphasised later by David Crawford of Kitt Peak National Observatory, we are currently doing too much of what he calls "second-class first-class astronomy" whereby astronomers with excellent programmes fail to complete them satisfactorily on large telescopes because of the inherent inflexibility of those instruments.

Due to the constraints of conventional large telescopes imposed largely by scheduling arrangements, based on the presence of human observers, many important branches of astrophysics are poorly addressed at present. Of particular note are four areas: (i) Rapid response to Targets of Opportunity, for example the outbursts of novae and supernovae. These unpredictable phenomena are a severe test of a conventional observatory's ability to override scheduled observers, and then to provide adequate follow-up. (ii) Regular and frequent long-term monitoring of variable objects in projects ranging from the determination of asteroid rotation rates, through studies of complete outburst cycles of cataclysmic variables, to quasar variability (including gravitational lensing on weeks to months timescales). (iii) Multi-frequency campaigns, particularly in relation to optical data which complement spacecraft observations at other wavelengths. For example over 60 per cent of the programmes proposed for the X-ray satellite EXOSAT required ground-based observations to maximise the scientific return. Of these only 25 per cent were successful in gaining such coverage, mainly due to the inflexibility of ground-based observatories. (iv) Finally, robotic telescopes are ideally suited to conducting large-scale, routine tasks, such as fundamental calibration work, plus the initial searches for objects such as supernovae and earth-crossing asteroids.

An added advantage of these telescopes is, of course, their cost-effectiveness. This is highlighted by Greg Henry and Doug Hall in this volume, who point out that their robotic telescope provides data at under one-tenth the cost per photometric point of that obtained on a conventional telescope of the same aperture.

2. The history of the subject

In John Baruch's paper, which reviews the current status of automated and robotic telescopes in Europe and provides a useful guide to definitions from "automatic" to "robotic", mention is made of Bart J. Bok's call at the 1953 Lowell Observatory Conference for a small automated monitoring telescope to provide extinction information at major observatories. By the late 1950s, E. S. Perfect was performing pioneering work automating the observing procedure for the Zenith Tube at the Royal Greenwich Observatory in England. The current Carlsberg Automatic Meridian Circle (CAMC) sited on La Palma in the Canaries is a descendant of this instrument and has the capability of performing its tasks fully robotically.

Much of the early work on automation of telescopes occurred, however, in the United States (see, for example, Genet 1986). The first of these automatic photometric telescopes (APTs as they became known) were developed by the University of Wisconsin and the Space Division of KPNO as a first step in their moves to provide telescopes in space.

The Wisconsin 8-inch telescope, situated at the Washburn Observatory Pine Bluff Station, was the brainchild of Arthur Code and was controlled by a DEC PDP-8 minicomputer. The optical system was identical to that of the Orbiting Astronomical Observatory (OAO) programme of space telescopes. The photometer was situated at prime focus and the telescope was used as an extinction monitor. It is a salutary point to note that its complex tasks were performed with only 4K of RAM. This telescope operated well for several years until it was finally shut down after students picnicking in the observatory grounds found that pouring beer through the rain sensor of the instrument caused the roof to slam shut!

The KPNO project aimed to automate a conventional 50-inch telescope and was ambitious for its day, including as it did weather sensors to open and close the dome. Due to the complexity of the task and the inadequacy of the computer power available they were lucky to keep the computers and other equipment going long enough to obtain one light curve of even a short period variable. The project ended with the telescope reverting back to manual operation.

The real push towards robotic observatory operation came with the application of micro-computers as opposed to mini-computers to the subject by David Skillman, an amateur astronomer working in Maryland. In 1979 his 12-inch automated telescope saw first light. This still required manual start-up at the beginning of the night and the degree of automatic response to rain etc. during the night was limited. The main application to which Skillman turned his telescope was observations of eclipsing binaries (Skillman 1981).

The next significant stride forward was made primarily due to Lou Boyd, an engineer and dedicated amateur astronomer who is the first to admit that he has never performed manual photometry. We are, however, fortunate that he is part of a network of amateur and professional astronomers centred around a fellow enthusiast, Russ Genet, who collectively appreciated the advantages of automated observation. Boyd set out *ab initio* to design and build a fully automated photometric system. This saw first light in November 1983. As pointed out by Hall and Henry later in this volume, this was indeed a watershed in the development of robotic observatories not least because this instrument was tackling in software the vexed problem of automated scheduling of observations. This is the key to efficient and effective observing.

Subsequent years have seen a growing participation by professional astronomers. In the United States the prime mover from the professional community is recognised as being Doug Hall whose main interest has been in RS CVn binaries. In this volume we see the fruits of his work with the Vanderbilt-Tennessee State 16-inch APT on Mount Hopkins and the successful efforts he and his collaborators later made to improve the accuracy of the photometry performed by this instrument.

The later 1980s saw the establishment of the APT service based on Mount Hopkins (some of the results of which are demonstrated in Don Fernie's contribution later in this book) and the founding of Autoscope as a manufacturer and provider of automated telescopes. Further adaptation and development of APTs for CCD imaging and photometry by the Berkeley group are of particular note (e.g., Richmond et al. 1992). Developments elsewhere in the world were also occurring, for example the automation of the Danish 20-inch telescope at the European Southern Observatory, and now robotic telescopes can be found in operation on every continent, including Antarctica (Chen and Wood 1991).

3. In these pages

When plans for IAU Colloquium 136 on Stellar Photometry were being drawn up in 1992 it was felt that it would be appropriate to hold a more informal meeting to discuss the impact of automation on the art and science of photometry. This book is based around the resulting workshop which was held in the ancient town of Kilkenny in southern Ireland in July of that year. The workshop attracted 27 participants from 13 countries. Initially, our ambition stretched as far as publishing the proceedings privately, however, it was subsequently recognised that the contents of the workshop in fact merited a wider audience due to the topicality of the subject and breadth and depth of the issues covered. Thus I am grateful to all the authors who developed and updated their contributions to September 1994 to provide the volume on the topic of robotic observatories that is presented here.

The book itself is divided into four parts. In the first of these, automated and robotic observatories, both extant and planned, are surveyed in approximate order of completion. Details are given of projects from those such as the Bradford robotic telescope, which is based on a ground-up design, to those of Fastnet in Ireland and Kotipu Place Observatory (New Zealand) where developments centre around converting commercially available small telescopes to automated observation for both astronomical research and science education. These illustrate the breadth of such ventures which are now at varying stages of development across the world. The most advanced employ CCDs for both target acquisition and data gathering. The result is that 1-metre class robotic telescopes are now aiming to observe beyond the $V=20$ boundary under less than perfectly photometric conditions. The wealth of observational programmes in which we will then be able to indulge is addressed in more detail in the Editor's Summary to this volume.

Part II presents details of observational results from automated or robotic telescopes. Here, for example, we see how improvements to drives, computers, control and instrumentation lead to increasingly accurate photometry, and that essential attribute of such telescopes—reliability of the data. The vast majority of the work currently being undertaken, as evidenced in this section and elsewhere in these pages, is stellar in nature and confined to our Galaxy. The paper on starspot lifetimes illustrates very well how homogeneous, large data sets, as generated by robotic telescopes, can be invaluable in helping to progress our understanding of a variety of astrophysical phenomena. Later in the book, particularly in Ian van Breda's contribution on AGN, it becomes clear that the near future holds the prospect of extragalactic astronomy reaping perhaps even greater rewards from robotic telescope operation.

The next stage in the exploitation of the potential of robotic telescopes, and the theme of Part III, is the development of networks of such instruments. These would give, for example, uninterrupted light curves of duration far longer than isolated telescopes observing at some fixed longitude can possibly provide. The potential of networks of observatories has already been demonstrated by, for example, the so-called "Whole Earth Telescope" which focused on observations of the oscillations of white dwarf stars (Nather 1991). Underpinning networks of telescopes are computing networks, such as those developed in Bulgaria and described here by Delchev and Tsvetkov. Anyone formerly unconvinced of the utility of small telescopes and how science emanating from them complements that of larger telescopes need look no further than the contribution by David Crawford.

The principles of networking exist in various forms in the robotic telescope community. In the last part of this book, as elsewhere, the spirit of co-operation and mutual support in which this endeavour moves forward is further reinforced. Throughout this part (which has the theme of technology and ideas) offers of, and requests for, help in various aspects of the subject are given. Here, however, some of the fundamental challenges of millimagnitude precision photometry emanating from networks of robotic telescopes are emphasised. In short, this final section aims to serve as a pointer to future directions for the subject which the concluding Editor's Summary develops further.

Finally, several individuals and organisations deserve thanks for their invaluable efforts. Ian Elliott and Russell M. Genet were instrumental in bringing together all the astronomers who took part in the workshop and acted as joint chairmen of the event. Generous financial support was forthcoming from Aer Lingus and Fred Hanna Limited. The local organising committee of IAU Colloquium 136 and the staff of the Dunsink Observatory supported the event in many ways. Thanks are also due to Dorothy Elliott and Andrew Elliott for helping with a diverse set of arrangements during the workshop. I am grateful to Clive Horwood and John Mason of Praxis Publishing who recognised that the proceedings of the workshop merited a wider audience and have provided invaluable guidance in bringing this project to fruition. Finally, without the tireless efforts of André Brabin in typing and formatting the text through several iterations, this book, unlike so many robotic telescopes currently to be found around the world, would not have seen its first light.

References:

Chen, K.Y. and Wood, F.B., 1991, in: *Robotic Observatories: Present and Future*, eds. S. Baliunas and J.L. Richard, Fairborn Observatory Press, p. 97.
Genet R., 1986, in: *Automatic Photometric Telescopes*, eds. D.S. Hall, R.M. Genet, and B.L. Thurston, Fairborn Press, p. 1.
Nather, R.E., 1991, in: *Robotic Observatories: Present and Future*, eds. S. Baliunas and J.L. Richard, Fairborn Observatory Press, p. 287.
Richmond, M.W., Treffers, R.R. and Filippenko A.V., 1992, in: *Robotic Telescopes in the 1990's*, ed. A.V. Filippenko, ASP Conf. Series, vol. 34, p. 105.
Skillman, D.R., 1981, Sky and Telescope, **61**, 71.

PART I

ROBOTIC OBSERVATORIES IN OPERATION, UNDER DEVELOPMENT AND PLANNED

A review of European developments in automated and robotic telescopes for photometry

J.E.F. Baruch

Department of Electrical and Electronic Engineering,
University of Bradford, Bradford BD7 1DP, UK.

Abstract

The hierarchy of telescope automation from automatic to fully robotic instruments is explored and defined. A review of developments in these areas shows that there is now a high level of activity and interest across Europe.

1. Introduction

In his opening address to the Kilkenny Workshop, Professor Wayman referred to the pioneering work of E.S. Perfect who developed an automated observing procedure for the Zenith Tube at the RGO (Perfect 1959). Although this was referred to as automation at the time, it comprised an automated routine for punching the transit time of stars onto cards to be processed at a later date by computer. An astronomer was still required to start off the routine for each star. This would hardly be called automatic nowadays. Today the RGO operates an automated transit circle which can operate robotically by itself but normally is switched on each night by the duty astronomer. It was originally described by Fogh-Olsen and Helmer (1978) and it has evolved over the years to facilitate complete robotic operation (Helmer and Morrison 1985, Helmer et al. 1991).

There is a history to the development of telescopes and their instrumentation which has followed a path that has for many years been called automation—sometimes, as above, where the meaning of the word has changed, and sometimes not. In 1953 at the Lowell Observatory Conference on Astronomical Photometry, Bart J. Bok called for "a small automatic monitoring telescope that would provide at all times exact extinction information" (Bok 1955). I believe Bok meant what we today would call an automatic telescope. What is currently meant by robotic and automated? Is it worthwhile trying to standardise terms and being more precise in their meaning?

In my view the standardisation of terms would allow the problems of the individual systems to be more focused and better addressed. The different systems have different applications and precise understanding of these applications will encourage collaboration and the development of common solutions to the problems. In this way the many applications of different levels of automation can be better directed and more effective.

2. Definitions

I will therefore give an overview of the terminology before moving on to review the current state of developments in Europe.

2.1 AUTOMATIC TELESCOPES

Automatic telescopes parallel the early industrial robot arm developments such as the programmable paint sprayer or the pick and place machine that is led through a series of movements as a form of training and then left to repeat the movements ad nauseam, only controlled by a sensor that determines whether the pieces to be picked or painted are present. In the case of the telescope, the star, its comparators, standards and sky are acquired by the human observer and then the telescope is left to monitor the object as it transits the sky. Such a system often has a wide aperture and sensors to look for dawn or precipitation. It continues to monitor the object until dawn or adverse weather conditions force an automatic shutdown.

Such telescopes are ideal for continuous monitoring. This currently includes the bright relatively rapid variables, the much more demanding observations used for stellar seismological studies, and the potentially chaotic behaviour of accretion disc decay onto condensed objects. Many systems for continuous observing have been developed by amateurs to enable them to obtain data with the minimum sleep-loss. The telescopes are exclusively equatorial mounts which are well set-up and able to track stellar objects with a sidereal clock. They have a relatively large aperture photomultiplier photometer. Professional astronomers are interested in the significant increase in photometric precision that automatic systems can provide when the human interference has been removed (Sterken and Manfroid 1991, Young et al. 1991).

The main current limitation of such telescopes is one of magnitude. At about $V = 12$ there is a confusion barrier (Baruch 1992). This is the point where the probability of having more than one star of this magnitude in the field of view approaches 50% and normal aperture photometric observations are precluded. A training approach to comparative photometry using CCDs is possible for fainter objects. The training is applied to the data reduction process for each field. When trained, the system is left to monitor the object and build up the data set. There are a number of larger telescopes that could be used in this mode of operation, but as far as I am aware the control loop has always included a human observer.

2.2 AUTOMATED TELESCOPES

Automated telescopes are telescopes able to find stars and centre their photometers on the target stars. They can follow a preloaded observing sequence and move around the whole sky. Such telescopes are ideal for service photometry. These telescopes imitate the developing flexibility of industrial robots which are able to mix a number of tasks and respond through their software programming to complex requirements. They require a significant leap in the sophistication of the telescope control system, drives and encoders.

Many large professional telescopes operate in this manner for photometry. The astronomer sits in the control room. The main task of the astronomer is to confirm that the correct star has been found and lock-on the star tracker. He/she also monitors the data output and checks the weather. For bright stars there is little for the astronomer to do apart from the routines at the start and end of the observing run. These are not difficult to automate. There are a number of automated systems which work well with bright stars. They are a natural extension to the automatic mode but require a telescope that is reasonably well set-up using one of the many telescope pointing software packages. One very good one is available on STARLINK (Wallace 1987). The key to automating lies in the drive systems and the encoders. If the encoders can be read automatically and there is a minimal amount of backlash in the drives then there is no reason why such a telescope cannot be automated.

Many satellites and radio telescopes operate as automated systems. They have excellent pointing and tracking capability. They receive their observing programme every 12 or 24 hours for the next observing period and return the raw data. In a few cases this is partly reduced in real time. For automated data reduction for faint objects it is necessary to produce pattern recognition parameters and atlases that can be universally applied to enable observers to check that the target object has been observed.

2.3 REMOTE TELESCOPES

Teleoperation is an important contribution that robots can make to operations in locations that for some reason are difficult places for human beings to reach. In the depths of the ocean, under oil rigs, on the surface of the moon, in nuclear reactors and (until recently) on the streets of Northern Ireland (defusing bombs) are all places where the remote operation of robots is effective. In astronomy there was a fashion in the mid-eighties to move into remote operations. It was argued that there would be considerable savings in air fares with the accompanying convenience of either observing from your office computer or from a national observing centre.

Since the earliest proposals by Robinson et al. (1982) a number of such systems have been provided at major observatories, e.g., from Garching in Germany to the telescopes in Chile for the European Southern Observatory, and from the top of Mauna Kea to sea level to save astronomers the discomfort of working at an altitude of 4,000 metres. In the USA the Apache Point Telescope has been built in New Mexico by a consortium of colleges specifically to provide their students with remote observing facilities on a large (2+ metre) telescope.

For all these systems there are a number of potentially useful additions. For robotic systems most such additions are essential. They include an automatic star tracker to correct the tracking, conditions monitors such as cloud, wind, dew and rain sensors to add to light sensors and provide more security for the telescope and its instruments. Apart from the availability of commercial trackers (e.g., Schwartz 1989) and a profusion of commercial weather stations for wind speed and temperature, some of which will interface to a personal computer, this still leaves a significant number of problems for astronomers who wish to automate. Various groups are tackling these problems but there seems to be little collaboration in these "peripheral" areas. The author has not been able to find any contributions to the relevant astronomy conferences over the past five years.

2.4 ROBOTIC TELESCOPES

Robotic telescopes observe autonomously. For normal operation the robotic telescope with its environmental monitoring system requires only electricity and a communications link to the site. Observing requests are sent to the telescope. Each is allocated a priority and the telescope schedules itself according to the priority of the requests and a number of other considerations, one of which is the observing conditions. The telescope finds the objects, produces a pattern recognition file to provide assurance that it has the right object. It takes the observations requested, reduces the data and returns the reduced data to the astronomer who requested the observations.

The key difference here is the complete autonomy of the system. The difficult developments are the systems which monitor the environment to ensure that the telescope can be operated effectively and safely, methods for monitoring its own operational efficiency and the development of indices associated with the data which give the astronomer confidence in the data set. The limitations of such systems are the large computing resource and intelligence that is required for all, including the simplest, operations.

Effective environmental monitors are essential for the operation of robotic telescopes. Bugs, birds, dust and small animals can play havoc with systems without a high degree of duplication and without the intelligence to monitor the operation of each sensor. Effective data quality indices are essential for robotic telescopes to win the confidence of the astronomical community. Without these use will be limited to a small band of enthusiasts.

The best known of the robotic telescope enthusiasts are the group centred around Russ Genet and his collaborators, originally in Phoenix, Arizona, and now in Fort Collins, Colorado. Their Automatic Photometric Telescopes, supplied by the Autoscope Corporation, are essentially a commercial operation. In Europe many observatories have been moving along these lines, linking their developments into the other work of the observatory and keeping quiet about their progress. Most of the projects in Europe centre around small (< 1-metre aperture) telescopes and for this reason do not have the high profile of the large telescope developments. However, they are confronting the next challenge in robotic astronomy which is to go faint and work at magnitudes of 16-plus.

3. A European tour

Europe has not been slow in rising to this challenge. A key component of the VLT European Southern Observatory Telescope to consist of four 8-metre mirrors will be its ability to operate as four separate automated or remote telescopes.

3.1 SPAIN

The Spanish Instituto de Astrofisicas Canarias has a developing interest in automated instrumentation to support the large collection of telescopes on the Islands of Tenerife and La Palma. Current developments are linked to automatic instrument developments in other fields such as medicine and robotics. They have a current telescope building and modernisation programme with the objective of producing instruments that can operate remotely, robotically and automatically. They have an active development programme for telescope finding and tracking, telescope scheduling and image processing in crowded fields. These are supported by general developments in intelligent control systems and a telecommunications network between the observatories on Tenerife, La Palma and the mainland.

3.2 FRANCE

In France there has been interest in all levels of automation over a number of years. Dr. J. M. LeContel and others converted a 60-cm telescope into an automated telescope in the early 1980s to produce a facility able to observe pre-defined sequences of variable stars. This telescope has been operating at the Sierra Nevada Observatory in Spain for the last 10 years. On the basis of this experience they proposed the installation of a completely robotic telescope at the South Pole to support their interests in stellar seismology and similar long-term projects. They are currently involved in building automated multichannel photometers to work with such telescopes. Other current French proposals include a group of three 1-metre robotic telescopes operating in the Atlas Mountains of Morocco linked by satellite to France, the development of automated interferometers, and a proposal to collaborate with the Saudi Arabians to build robotic telescopes. (See also the chapter by Querci et al. in Part III of this volume.)

3.3 BELGIUM

The Belgian group led by Chris Sterken has been actively involved in the development of automated and robotic systems for many years. They have experience of using standard photometric systems and telescopes operated by people for the long-term monitoring of variables. It was this experience that led them to point out that high-precision consistent photometry will be performed most effectively by machines. Robotic telescopes will produce a revolution in the quality of photometric data (Sterken and Manfroid 1992).

3.4 UNITED KINGDOM

There has been an Automated Monitoring Telescope Panel working to generate funds for a UK robotic telescope. Their main success has been a survey of the UK astronomical community which produced proposals for over 70 observing programmes for a 1-metre automated telescope and would have kept such a telescope busy for many years.

There is a considerable amateur effort in the UK with a number of automatic telescopes in operation. The leader in the field was Els with his automatic observatory. A number of university observatories have built automatic systems for the use of students, e.g. Birmingham (Elliott and Eyles 1987). The major observatories with their telescopes in Hawaii and the Canaries have automatic capability for the simplest operations on most of the telescopes and remote capability for some operations.

The major robotic development is at the University of Bradford in the Electrical Engineering Department where we are building a telescope as a technology test bed for robotic operations which will be linked into the academic communications network JANET. It is designed for robotic operations with four instrument locations. The intention is to place it in the Canaries when it has been shown to be effective in the UK.

3.5 DENMARK

The Danes have been involved with robotic operations for many years. The first was the Carlsberg Automatic Meridian Circle (CAMC; Fogh-Olsen and Helmer 1978, Helmer et al. 1991) now operating on the island of La Palma. They have also automated the Danish 50-cm telescope operating at the European Southern Observatory site at La Silla in Chile with a Strömgren six-channel photometer (Nielsen Florentin et al. 1987). This is an automatic telescope where the observer programmes the observing sequence for each night and then leaves the telescope to follow the programme. The telescope can also be operated remotely from Europe.

3.6 ITALY

There are a number of programmes to produce automated and robotic systems. At Asiago in the north there is a developing programme to automate the telescopes at their newer site higher up the mountain. The Catania group on the island of Sicily have purchased an 80-cm APT from Autoscope in the USA which they have installed at an altitude of 2000 metres on Mount Etna. There is an active programme of building automatic instruments to provide optical monitors on space craft, particularly to work with the JET-X experiment on the Soviet Spectrum-X-Gamma satellite (Antonello et al. 1990).

The main Italian group developing robotic instrumentation is centred in Naples around Mancini and Longo at the Osservatorio Astronomica di Capodimonte (OAC). They have already built three automatic telescopes: a 40-cm operating at OAC, a second 30-cm at Castelgrande and a third at the Observatory of Perugia—a 40-cm telescope with a CCD. They are currently connecting the telescopes to the Italian national communications network ASTRONET for remote as well as automatic operation. They are also building a new automatic 1.5-metre telescope for the Observatory of Castelgrande. The objective is to have all these telescopes running in automated and robotic modes linked to national data centres.

3.7 BULGARIA

The Bulgarians have automated a 60-cm telescope at Belogradchik. They have incorporated a single-channel photon counting photometer. They are currently working on linking the telescope into the developing Bulgarian academic communications network. They are also automating a second telescope to obtain simultaneous multicolour observations of very rapid stellar flares.

3.8 HUNGARY

There is a consistent long-term effort in Hungary to move into robotic systems. There are two centres. The Sambothay Gothard Observatory are working on developing networks to facilitate remote operations for their telescopes in parallel with their telescope automation programme. The Konkoly Observatory has a programme to automate their 1-metre telescope.

3.9 OTHERS

There are many other unpublished developments proceeding in Europe at present. The ESO VLT programme is bringing in many groups to support its automated operation. There are also groups in Romania, Russia, the Czech Republic and the Ukraine who are working towards automated systems. Overall, therefore there is a very high level of interest and activity in Europe in this area which is developing very rapidly.

References:

Antonello, E., Citterio, O., Mazzoleni, F., Mariani, A., Pili, P. and Lombardi, P., 1990, *Proceedings of the SPIE 1990 Symposium on Astronomical Telescopes and Instrumentation for the 21st Century*, Tucson, Arizona, February 11—17.
Baruch, J.E.F., 1992, in: *Variable Star Research an International Perspective,* eds. Percy J.R., Mattei, J.A. and Sterken, C., Cambridge University Press, p. 109.
Bok, B.J., 1955, *Astron. J.*, **60**, 32.
Elliott, K.H. and Eyles, C.J., 1987, *Journal of the British Interplanetary Society*, **40**, 195.

Fogh-Olsen, H.J. and Helmer, L., 1978, *IAU Colloquium No. 48*, Institute of Astronomy, University Observatory Vienna, p. 219.
Helmer, L. and Morrison, L.V., 1985, *Vistas in Astronomy*, **28,** 505.
Helmer, L., Fabricius, C. and Morrison, L.V., 1991, *Experimental Astronomy*, **2**, 85.
Neilsen Florentin, R., Norregaard, P. and Olsen, E.H., 1987, *The ESO Messenger*, No. 50, p. 45.
Perfect, D.S., 1959, *Occasional Notes of the RGO*, **3**, No. 21, p. 223.
Robinson, W., Schechter, P. and Janes, C., 1982, *Advanced Technology Optical Telescopes, SPIE Proceedings*, **332**.
Schwartz, R., 1989, *IAPPP Communications*, No. 38, p. 34.
Sterken, C. and Manfroid, J., 1991, *The ESO Messenger*, No. 63, p. 80.
Sterken, C. and Manfroid, J., 1992, in: *Precision Photometry*, ed. K.A., Jones., L. Davis Press, p. 335.
Wallace, P.A., 1987, *Pointing and Tracking Algorithms for the Keck 10 metre Telescope. The Ninth Santa Cruz Summer Workshop in Astronomy and Astrophysics*, Lick Observatory, Springer-Verlag, Berlin.
Young, A.T. et al. 1991, *PASP*, **103**, p. 221.

A progress report on the Sutherland automatic telescope

D. Kilkenny

South African Astronomical Observatory, Observatory 7935, South Africa

Abstract

A report is given of progress on the APT currently being constructed for the Sutherland site of the South African Astronomical Observatory by the SAAO and the Universities of Cape Town (UCT) and South Africa (UNISA)

1. Introduction

The Automatic Photoelectric Telescope (APT) planned for the Sutherland site of the South African Astronomical Observatory (SAAO) is being jointly funded by the SAAO, the Universities of Cape Town (UCT) and South Africa (UNISA) and the Foundation for Research Development. Most of the electronic and mechanical work has been done at the SAAO and much of the software written at UCT. The optical components are being figured by the Optical Engineering Programme of the Council for Scientific and Industrial Research (CSR).

A review of the progress on the Sutherland APT was precisely given by Kilkenny (1990; see also Warner 1991, Kilkenny 1992).

2. Housing

The building is essentially complete and the Ash-Dome is fitted. Power rails and slip-rings for the dome are installed and the lower flap has been motorised (the "roll-over" shutter has an extra 5° of travel to facilitate observations near the zenith, but this means the lower flap could interfere with low-altitude observations. It is planned that the flap will be lowered for such observations but raised otherwise, to give additional protection). Some minor interior fittings have still to be done; for example, we plan to install a fan system to reduce diurnal temperature variations in the dome.

3. Telescope

The telescope is being built, effectively under licence, to the basic Autoscope design (e.g. Genet et al. 1989) and most of the mechanical structure has now been completed (see Fig. 1). At the time of writing (September 1994), we were still awaiting confirmation of the final optical parameters before assembly. Software for the telescope control is written, but not yet tested.

The Hextek blank for the primary mirror is currently being figured at the CSIR's Optical Engineering Programme in Pretoria. Grinding of the primary is complete and polishing underway; the secondary has been ground, but will not be figured finally until the primary is completed. It is hoped to take delivery of the optical system in late 1994. We plan to carry out preliminary testing of the setting and guiding of the telescope and the optical performance in Cape Town before moving the telescope to the Sutherland site.

The construction of the telescope and photometer have been carried out under the direction of Digby Ellis (SAAO) who also designed the photometer itself.

Fig. 1: The main telescope structure of the SAAO APT.

4. Photometer and CCD

At present, we envisage a fairly conventional photoelectric photometer for the APT. This is to simplify certain aspects of construction and data handling, though with the continuing improvement of the characteristics of CCDs and the data-handling software, it might prove feasible to replace the photomultiplier system with a CCD-based photometer at a later date. We therefore plan to test and run the system initially with a tube-based photometer. We will, however, use a commercial-quality CCD for field acquisition and offset guiding.

Work on the filter and aperture wheel units for the photometer is completed, and a Peltier effect thermoelectric cooler has been purchased for the photomultiplier tube. It is planned to use a Ga-As tube to permit photometry in the UBVRI wavelength range.

Several CCD systems have been designed and constructed at SAAO, based on the RGO/RAL "Merlin" transputer/controller (Waltham et al 1990); one is already in use on the 1.9-m telescope for acquisition/guiding and one will shortly be available for the APT. The CCD is an EEV 578 × 386 chip used in the frame-transfer mode and cooled to about $-45°C$. The cryostat was designed by Ian Glass (SAAO) and uses a three-stage Marlowe Industries thermoelectric cooler.

Our original plans for the acquisition and guiding included using software by Hine and Nather (private communication) and, while we still plan to test this, Greg Cox (UCT) has written completely new point-pattern-recognition software for the Sutherland APT. This is based on matching similar triangles in a given pattern of points, using the largest and smallest angles of the n[comb]3 triangles which can be constructed between n points (star positions). The method is scale, rotation and translation invariant and is relatively insensitive to adding random points or removing sections of the basic pattern. So far, the software works well on artificially generated data and on CCD frames taken with other instruments, though full testing has not yet been done.

5. Electronics

A large part of the electronics is already in existence, since some of the boards are duplicates of units used in other instruments—the time, display and photometer interface boards, for example. An SAAO filter control card designed by Dave Carter and residing in the transputer system will control the photometer filter and aperture wheels. The telescope (RA and Dec.) control system, based on a printed circuit board purchased from Autoscope, is essentially complete. Work on the design of the electronics for dome control, focus control and moving mirror control is currently in progress.

A number of different designs by Guy Woodhouse (SAAO) for the overall electronics system have been considered, but the original concept of two PCs, one to control the photometer and the other to act as "observer" (programme organisation, telescope setting, acquisition/guiding, dome control, etc.), still seems the simplest. The main departure from the original concept (see Kilkenny 1990) is that some of the control functions will now be performed by the CCD controller. The PCs originally purchased to run the APT have now been replaced by faster "386" type machines.

6. Software

Code for the programme management, organising nightly programme star priorities and so on, has been written by Angela Jones (previously at UCT) and it is intended to use existing SAAO software for the photometer control (Balona 1988); this also provides a number of status checks, such as voltages, and filter wheel position. Pattern recognition software for acquisition and guiding and code for telescope control has been written by Greg Cox, as described above. A good deal remains to be done, principally in the software which will simulate the observer (telescope focus, aperture selection and preliminary quality control of the data, for example). Although it is intended that the APT will operate fully automatically, we do intend to make use of the fact that there is always at least one other telescope which is making photometric observations at the Sutherland site. Fibre-optic links (Geoff Evans SAAO) will be provided to each of the other domes so that the system can be monitored for any instrumental failure (e.g., voltage failure to tube, amplifier, telescope etc.) and technicians notified. Additionally, one (human) observer will probably have the ability to initiate or override some functions such as dome opening (start-up) or re-opening after bad weather.

References:

Balona, L., 1988, 'LUCY Version 3, *Photometer control and data acquisition—Observer's Manual, SAAO*, Cape Town.
Genet, R.M., Hayes, D.S., Epand, D.H., Boyd, L.J. and Keller, D.F., 1989, *Robotic Observatories*, p. 115, Autoscope Corp., Mesa, Arizona.
Kilkenny, D., 1990, in: *11th Annual Smithsonian/Fairborn/IAPPP Symposium on APTs*, p. 277.
Kilkenny, D., 1992, *South African Journal of Science*, **88**, 245.
Waltham, N.R., van Breda, I.G. and Newton, G.M., 1990, in: *Proc. SPIE Conf., Instrumentation in Astronomy VII*, **1235**, 328.
Warner, B., 1991, *Proc. 3rd New Zealand Conference on Photoelectric Photometry*, in: *Southern Stars*, **34**, No. 3, p. 4467.

Developments in automatising a Celestron-based APT

R. Hudson[1], G. Hudson[1] and E. Budding[2]

[1] *Kotipu Place Observatory, Wellington, New Zealand.*
[2] *Carter Observatory, Wellington, New Zealand.*

Abstract

Carter Observatory's Celestron C14 APT (Loudon et al. 1992) has been redeveloped and reinstalled at the privately owned Kotipu Place Observatory, in a Wellington suburb. Stages of this operation are reviewed, and examples of data acquired by the instrument presented.

1. Background

Motivations driving the APT concept in a New Zealand context were reviewed by Loudon et al. (1992). Among the more general arguments is that the extra-human dimension in automated telescope operation for routine observational tasks implies greater reliability, speed, and production: that advantage which has been used to marked effect in large quantity automobile assembly, for example.

A newer slant favouring low-cost APT development comes from considering the rise in use of the "home-based" personal computer (PC), when set against the practice of working with the large campus main-frame of a generation ago. Perhaps a parallel can be made with this if effective low-cost APTs can be produced in sufficient numbers, at least for certain kinds of routine observational task in astronomy. It is feasible that home or locality-based APTs will make genuine inroads into extending good-quality surveillance of variable stars.

Loudon et al. (1990) described the initial development and installation of a low-cost Celestron "Compustar"-based APT at the Kelburn site of Carter Observatory in Wellington. They argued that its performance would be significantly enhanced by relocation at a site with a better sky, and full access to the Southern hemisphere (which the installation at Carter Observatory in Kelburn could not provide). This has been effected with the transfer of the Celestron to the Kotipu Place Observatory—a much darker, sheltered location in the small village of Pukerua Bay.[1]

2. Setting up the Kotipu Place Observatory

Work began in 1989 with the transport of a 2.4-m fibreglass dome from a previously used Northland site to Pukerua Bay, some 35 km north of Wellington and 20 km south of Paraparaumu. The observatory site is about 100 m above sea level. Light pollution is limited, coming mainly from the public lighting of about a dozen streets. The main disadvantages are that the Number 1 state highway and main trunk railway pass through the hamlet. We sited the observatory as far away from both as reasonably possible to minimise

---

[1] This step was instigated by E. Budding, who, on behalf of Carter Observatory, is grateful for the accommodating and productive response from the Kotipu Place Observatory.

noise level, vibration and car headlights. Heavy freight trains cause noticeable vibrations when they pass through, but this occurs infrequently at night.

The observatory, in a two-storey building, has a good southern view except for some glow from Wellington city lights beyond intervening hills. The horizon altitude is around 2 degrees. The northern horizon over the house is about the same. The eastern horizon is about 6 degrees over the southern Tararuas, while to the west hills rise to about 12 degrees. Detailed plans are available on request. The telescope supporting pier rises 3.5 m above ground level. One and a half cubic metres of concrete were poured to form a firm foundation for this 36-cm pre-stressed concrete pipe, filled partly with reinforced concrete and topped up with sand. A plastic electricity conduit runs up its middle. The pier was later strengthened by extending the base a further 2 m below the initial foundation, so that pier vibrations are now undetectable within a second or two of a heavy train passage. The dome was mounted on a laminated wooden ring. A second ring was attached to the dome to give it strength and a smooth, even surface for its skateboard rollers. Seven double power points were fitted in the walls below the dome. These walls were then lined with 12-mm chipboard. By July 1991 all was ready for use.

After some discussions and trials, involving the present authors, it was decided in early 1992 to resite the Celestron C14 "Compustar" APT (Loudon et al. 1992). It is mechanically preferable to balance the telescope in order to reduce pressure on the worm and achieve a regular drive action, particularly in the microstepping mode. Resetting the worm is not too easy, as the whole telescope and RA drive unit require disassembling for adjustments to be made on a bench. A persistent trouble was that whenever the instrument was reset on the pier the shaft moved slightly out of true fitting, due to distortion caused by the extra weight of the replaced telescope. A good setting was eventually located.

In order to align the polar axis, we first followed a user-manual recommended procedure involving slewing to σ Oct and adjusting the mount to centre the star. One setting star will not remove the errors of both the "park" initial position and the axis misdirection, however, so we could not quickly converge to a consistent fixing in this way.

Using the Compustar's digital readout we were able to see the error in altitude by pointing to a star on the meridian and setting the declination read-out to the given value. A progressive correction procedure involved turning the telescope over (cf., e.g., Sidgewick, 1955, Sinnott 1978) and recentring the star by adjusting the axis elevation trims. The azimuth setting could then be satisfactorily improved using the user-manual procedure as a single correction. Large slews ($\sim 90°$) now bring any given star to within about 10' of the centre. Further fine-tuning may improve this.

We constructed a special short dew shield because the Schmidt—Cassegrain corrector plate frequently fogged up. Because the telescope needs to be able to pass through the forks for parking, and because the CCT may well move the telescope through the forks while slewing, the dew shield is restricted to about 75 mm. The shield has a ring of resistors emitting ~ 10 W to reduce dewing. This power may have to be increased, because fogging still occurs occasionally.

3. Installation of the APT

3.1 THE GIVEN CONFIGURATION

The C14, as purchased, was automated through a computer-controlled telescope (CCT) unit, a commercial front-end for the Celestron designed as an observing aid. Its basic

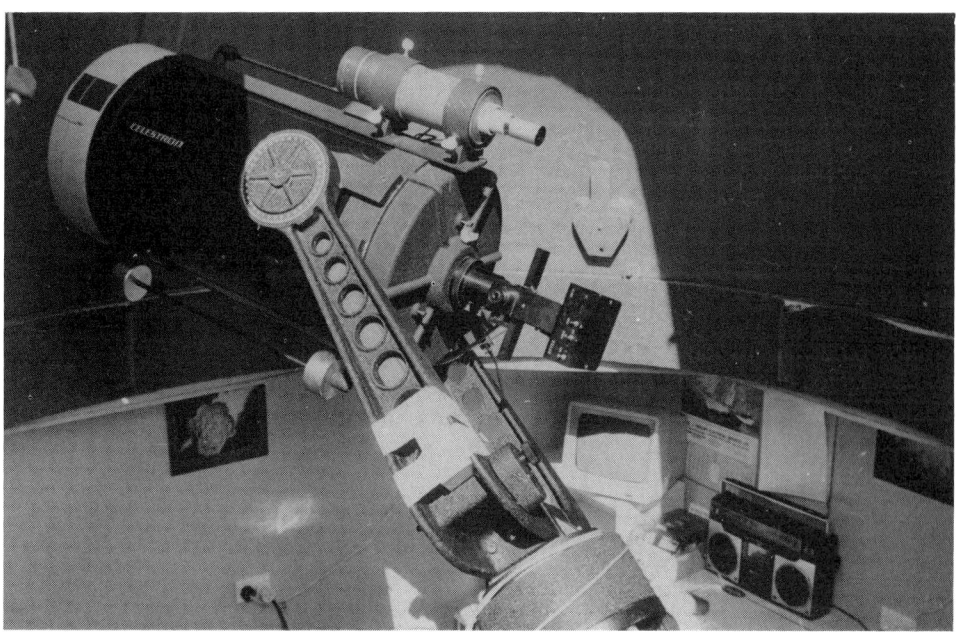

Fig. 1: The C14 APT installed at the Kotipu Place Observatory.

functions allow automatic positioning to any given equatorial co-ordinates (usually of objects extracted from internally stored catalogues via its key-pad) taking into account current time, latitude and precession. The telescope can be also moved manually via a plug-in joystick at either slew, set or guide speeds. Two pieces of electronic hardware were added: a high-current power supply to replace the heavy-duty driver battery—the system was designed for portability—and a joystick emulator to enable a PC-type microcomputer to move the telescope via the joystick socket. The CCT is an open-loop system, so the PC merely instructs to slew to given co-ordinates, without any check on what subsequently happens. The accuracy of setting relies on precise polar alignment of the telescope and initialising.

The photometer used is an Optec SSP 3a, a pin-diode device with support electronics to produce a square wave output, whose frequency depends on the light flux on the diode. A card in the PC provides additional frequency-counting circuitry and control of the filter-slide stepper motor. The photometer is the only source of feedback for control from the PC. This computer really should have additional information to establish that a located object is the right one. For practical purposes, however, the operator sets a "window" in which the photometer count must lie. Initial approximate settings then lead, almost inevitably (for bright stars), to the desired result. Window settings can vary for a particular target, depending, for example, on the level of moonlight.

The primary issue in the automation is getting the star into the aperture after its approximate setting by the CCT. Boyd's "HUNT" algorithm (Boyd and Genet, 1984) has been adapted to achieve this. It works well enough, but seems less time-efficient than human hunting with present control arrangements. Hunting is followed by centring the star in the aperture to ensure that the photometer is not missing light, and to enable a maximum effective measuring time. The centring process has so far been hindered by the C14 moving in somewhat uneven steps given identical commands. This appears related to frictional effects (increase in step size as kinetic friction takes over from static), as well as the low sampling rate of the joystick port not enabling the required positioning accuracy (Loudon et al. 1992).

3.2 DEVELOPMENTS SINCE KOTIPU PLACE INSTALLATION

A useful addition to the foregoing was a changeover box to allow selection of joystick or PC control of the CCT without reconnection. Manual intervention is still required during initialisation, and some back-up operations.

To eliminate certain manual actions a "front-end" for the PC control program was written, to enable easy access to databases like *Skycat 2000* and *NGC 2000*,[2] as well as support the positions of Solar System objects and a user catalogue. An intelligent initialising sequence was added to upgrade primitive actions existing in already written software. The program was written in Turbo Pascal, and it works well in slewing the telescope to set objects. It does not control photometry, however, for which a separate program was developed.

The first program of the new development in logging photometric data was simply to check the behaviour of a single star (flare star patrol). If the photometer count dropped below a certain threshold, the star was deemed "lost" and steps were taken to relocate it. This meant, in effect, invoking the HUNT routine. Two problems were encountered—the bigger was that the count from the re-found star was, more often than not, significantly different from what it had been just prior to star loss, due to drifts and ineffective centring. A new and better centring algorithm was installed, making use of a setting scan to carry out a "bisecting chords" procedure, and thereby bypassing the step-irregularity difficulty found previously (Loudon et al. 1992). The second problem was that this software was too dedicated, and offered no scope for other types of photometry.

At first the plan was to develop outwards from the flare star program. Later, however, we opted for complete redesign. The resulting program was developed over several weeks and currently has the features: (i) A hierarchy of observation information with data on variable, comparison, check, sky and other miscellaneous objects being grouped in a file for instant recall. Several such files can be grouped to form an observing list. Such lists are stored and recalled as necessary. (ii) Any single observation header is fully and independently definable. It includes: position of star, epoch of co-ordinates, which filter(s) to observe with, number of observations per filter, integration time per observation, the "found" threshold with clear glass filter, whether HUNT and/or centring is to be used. (iii) An automation sequence to observe a set with, e.g., 11,129 × 5 to observe the variable (1) three times (say UBV x 3), then observe the comparison (2) through the three filters once each, then the sky (9), then repeat the whole thing five times. (iv) Various global controls such as how to cope with "star not found", when to start and stop observing (other factors permitting); dealing with not being able to locate a star, which consists of specifying how many times it should try to search, then what to do should it repeatedly fail to find. It can skip the set and move to another, perhaps returning later, or close down if it fails to find other stars (usually because of cloud or the obstruction). The option is also present for "waking up", after a specified interval, to restart operations.

---

[2] Available from Sky Publishing Corporation.

4. Results

We show in Fig. 2 a montage of light curves of the bright δ Scuti-type variable ρ Pup, which have been observed with the APT. Comments have been made on these data elsewhere (Hudson et al. 1993). Although the star is quite bright, the possibility of frequent patrol at millimagnitude accuracy is a definite option.

5. Recent developments

A persistent problem has been that the PC, having instructed the CCT to move, is then "blind". This can cause possible cable fouling or difficulties with the timing of instructions. One solution is to add coarse position encoders to inform the computer of the approximate instantaneous position of the telescope. A reasonable extension, for when there is doubt about how a star will be approached, would be for the PC to slew the telescope in the sense programmed as most suitable, then use the CCT, with its higher precision, to fine position. An interim summary of developments currently considered can be made as follows: (a) Relating, or making compatible, our control software to ATIS or ATIS-plus format (Genet 1992). (b) Controlling the dome by the PC, to allow automatic monitoring of stars in any part of the sky, and follow them through longer periods. The shutter should similarly be controlled to protect the equipment from rain, in the case of untended operation. (c) Scanning weather instruments to allow the computer to make decisions about observing. (Full sequences to deal with cloud-over have not been fully tested yet.) (d) More effective demisting of the corrector plate. Efforts hitherto have only been partially successful on damp cold nights, when condensation can also occur on *inner* surfaces.

Other ideas for possible later work are the implementation of more remote control, firstly via a serial link to another computer in the adjoining house, and later via modem to selected users.

References and Bibliography:

Boyd, L. J. and Genet, R.M., 1984, in: *Microcomputers in Astronomy II*, ed. R.M. Genet and K.A. Genet, Fairborn Press, Ohio, (cf. also M. Trueblood and R.M. Genet, *Microcomputer Ctrl of Telescopes*, Willmann-Bell, 1985, p. 318).
Budding, E. and Trodahl, H.J., 1987, *Southern Stars*, **32**, 19.
Budding, E., 1991, in: *Third New Zealand Conference on Photoelectric Photometry*, ed. E. Budding and J. Richard, p. 61.
Genet, R.M., 1992, Circulated electronic mail (Genet@pegasus.edu).
Hudson, R., Hudson, G. and Budding, E., 1993, in: *Stellar Photometry, IAU Coll. No. 136* (Poster Papers), eds. Elliott, I. and Butler, C.J., Dublin Institute for Advanced Study, p. 107.
Kholopov, P.N., 1987, *General Catalogue of Variable Stars*, "Nauka", Moscow.
Loudon, M., Priestley, J. and Budding, E., 1990, *Southern Stars*, **34**, 1.
Loudon, M., Priestley, J. and Budding, E., 1992, *Proc. Joint Comm. Meeting on Automated Telescopes for Photometry and Imaging,* (IAU 21, Buenos Aires), ed. S. Adelman (in press).
Sidgewick, J.B., 1955, *Amateur Astronomer's Handbook*, Faber and Faber, London, Ch. 16.
Sinnott, R.W., 1978, *Sky and Tel.*, **55**, 78.

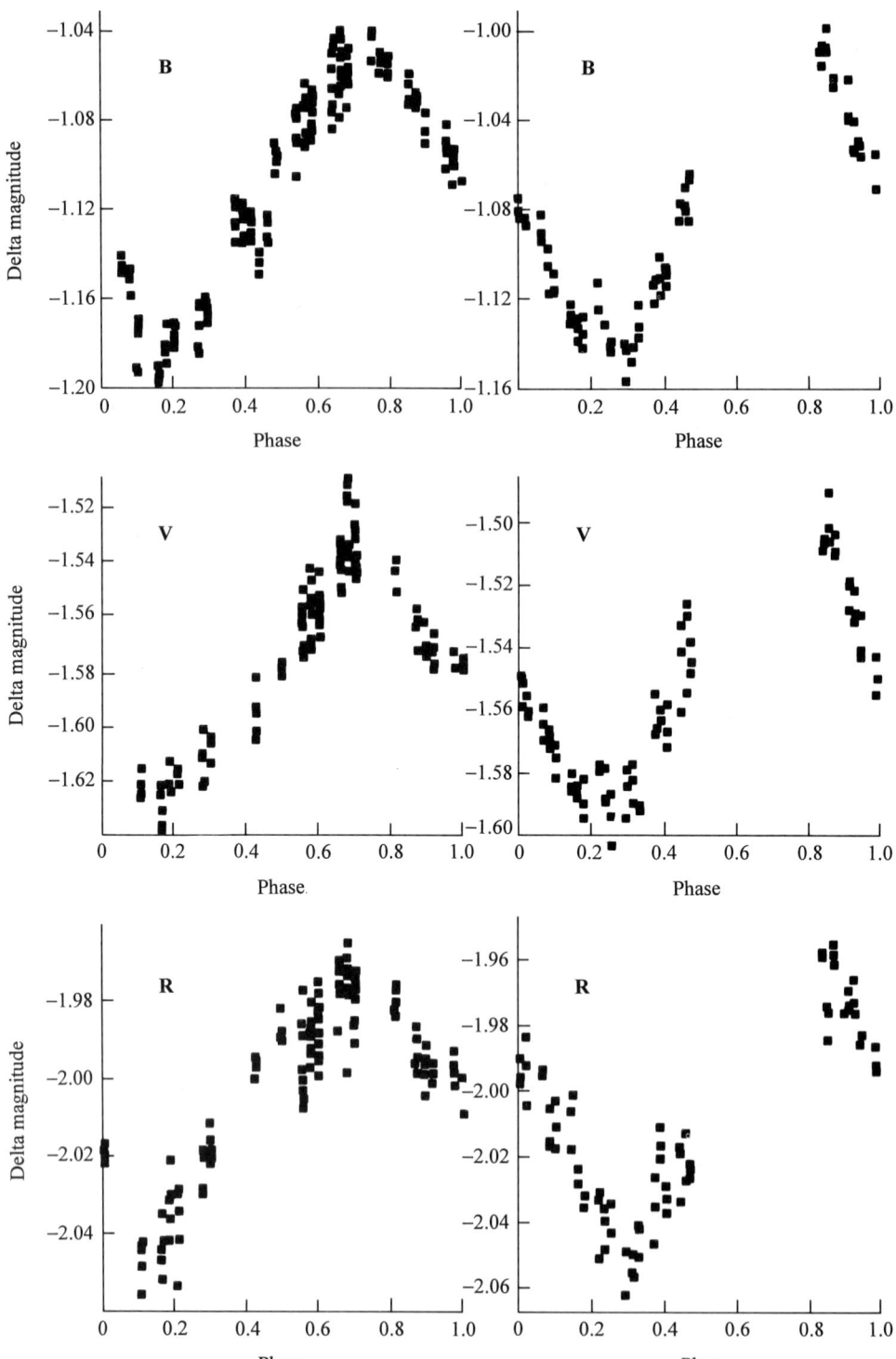

Fig. 2: Light curves of ρ Puppis.

The development of a robotic telescope system at Bradford

J.E.F. Baruch

*Department of Electrical Engineering, University of Bradford
Bradford BD7 1DP, UK.*

Abstract

The control system of the Bradford Robotic Telescope is outlined. It is shown how the telescope is modelled as a flexible imperfect system to produce a matrix net of corrections for the pointing in real time. The data flow for the production of the altitude and azimuth drives are discussed with a review of the sources of error and the expected precision.

1. Introduction

The Bradford Robotic Telescope is being constructed at the Bradford University Oxenhope Experimental Station, a relatively dark site in the Pennines about 20 km from Bradford and at an altitude of about 445 m. It can be argued that there are no good sites in the UK but the site is adequate to evaluate the system before transferring it to a good site. The telescope is designed to work in crowded fields with faint objects (i.e. fainter than m_v = 12). An outline observing programme is described by Baruch (1992).

Fig. 1: The Bradford Robotic Telescope in its building at Oxenhope.

The telescope is a 46-cm Newtonian reflector with an alt-az mounting and a cooled CCD detector. The CCD is a thinned GEC with 585 rows of 385 pixels. It has MPP dark current reduction and a UV coating. The telescope has four instrument stations, two for large or heavy instruments with the focal points taken through the altitude axes and two for light instruments, e.g., TV cameras, on the tube at right angles to the altitude axis.

The telescope is currently linked by modem to the control site at Bradford University which is linked into the international computer networks. The central controller is the master system responsible for the operation of the telescope, the evaluation of its performance and all the housekeeping. It also controls the communications and the scheduling of the system. A schematic diagram is given in Fig. 2.

The central controller is supported by a weather station. The weather station consists of multiple monitors for cloud, wind speed, wind direction, rain, snow, dew, ice, humidity, temperature, human presence and light. A lightning detector will be added soon. The telescope is in its own customised enclosure with environment sensors and a roll-off roof controlled by a fail-safe positive action drive system. The environment is continually monitored and an observing conditions file in the controller is continually updated to facilitate efficient and safe operation by the controller. These cover optimising the observing programme with respect to the local observing conditions. There are also measures to ameliorate dew conditions, partial close-down in the presence of people and heaters to limit the build-up of ice on the roof. The system is currently running through a rigorous programme to test the security of the weather station for robotic operation.

The image control and image processing system focuses the optics, controls the filters, provides stellar pattern recognition parameters for field confirmation and ascertains the observing conditions from an evaluation of the stellar image profiles. It evaluates the image quality and generates differential photometric data for the objects in the field. The basis of the image control system is the CCD controller supplied by Wright Instruments (Enfield, UK) and the image processing software PC Vista. The software is relatively versatile and designed to run on an IBM personal computer or clone.

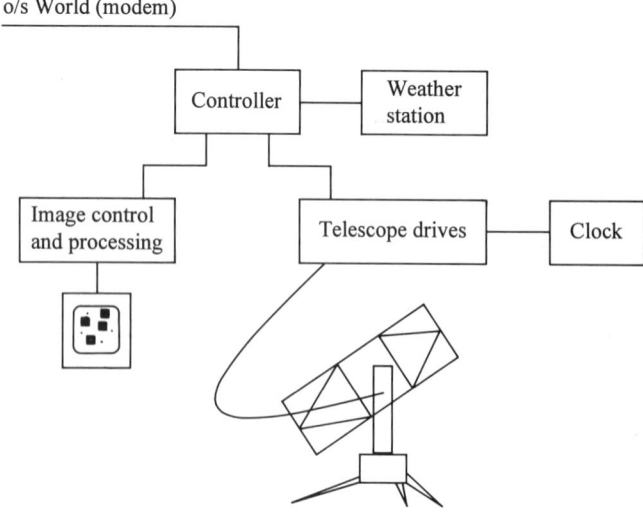

Fig. 2: Schematic of the system.

These data are reduced and compared with the observing request. When the data reach the precision required, or after a specified number of attempts, the results are returned to the requesting astronomer over the network. Photometric data can be supplemented by images.

2. The telescope control system

The telescope engineering drive system consists of four components:

- The stepper motors. At the lowest level a number of pulses or a pulse rate is fed to the stepper motors which are Compumotor A83/93 with a torque of 7 Nm. The motors require 25,000 steps to produce one revolution which is divided down by a direct drive system to give about 10,000,000 pulses per 360-degree rotation of either axis.

- The encoders are relative encoders with a precision of one arc second. This is degraded by a factor four by running them in quadrature. The stepper motors subdivide the encoder pulses for high precision. The encoders are supplemented by two fiducial markers on each axis.

- The drive system has an indexer drive system PC21 supplied by Compumotor which enables the velocity and acceleration to be defined along with the number of steps. In this way the position can be predicted.

- The image processing system allows the position of a star to be determined in the image plane which coupled with the clock driven STARLINK[1] software TPOINT provides a check on the tracking and pointing.

The system is supplemented with limit switches in hardware and software to ensure that the telescope position is always within the envelope of safe operation. Time is provided by a global positioning satellite system (GPS) accurate to about 50 ns when the full satellite complement is in orbit. It also gives a continuous readout of position and altitude.

The drive system is a direct friction drive using steel on steel. Considerable care has been taken to remove the effects of temperature and to ensure that the axes of the drive wheels are parallel, and parallel with the telescope axis and stiff enough to remain so under varying torque conditions.

Overlying the engineering drive system is the telescope model. The telescope structure is presumed to be flexible and the telescope and drive system axes are presumed to be non-central and non-perpendicular. The model has 17 parameters with the option of adding more second-order terms. Initially the model parameters are determined by driving the telescope around the sky and looking for consistencies in the pointing errors using the STARLINK TPOINT software. This procedure is repeated every night to provide precision assurance to the telescope monitoring system on a nightly basis. With the model parameters determined a matrix of corrections around the current pointing position can be produced. The resolution of the matrix declines as it moves away from the current position.

---

[1] STARLINK Project, DRAL, Chilton, Didcot, OX11 0QX, UK.

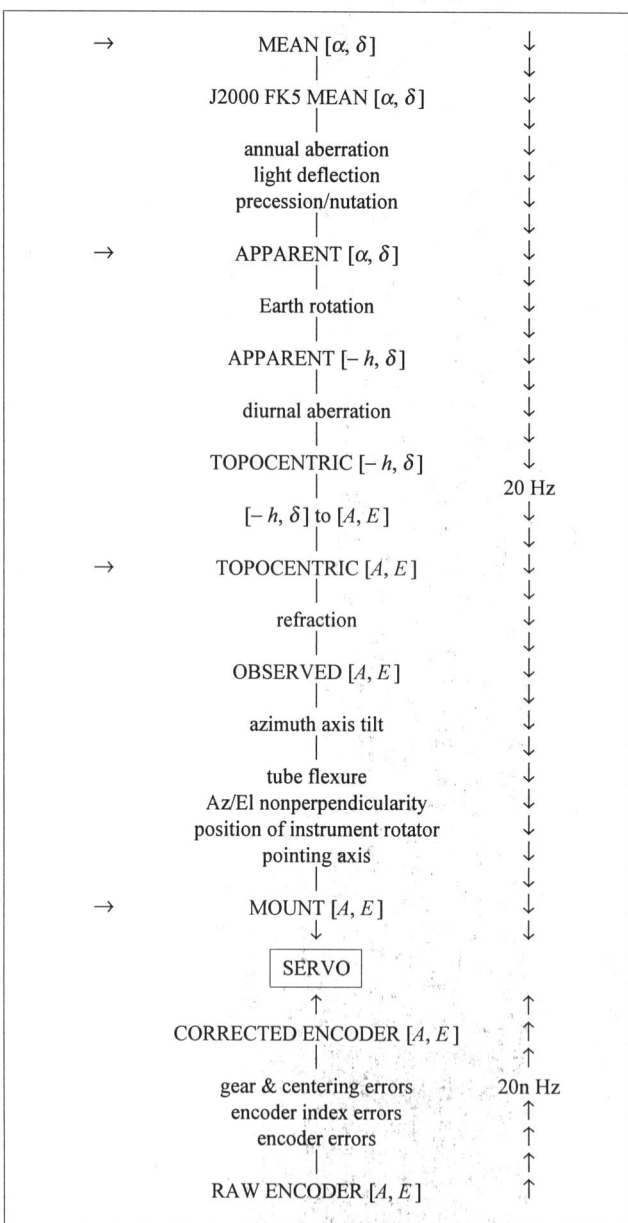

Fig. 3: The pointing flow. The set of transformations shown describes the relationship between the target position (one of those marked →) and the required telescope encoder readings. There are two major transformations: $[\alpha,\delta]$ to $[-h,\delta]$, and $[-h,\delta]$ to azimuth and elevation $[A,E]$. The others are all minor.

The inputs to the model are the velocity and distance in steps that the telescope is required to move in altitude and azimuth. The outputs of the model are the corrected movements incorporating the flexure in the telescope structure and the errors in the positioning of the axes and drives. The corrected drive instructions discussed above are derived from the model produced by Wallace (1987) for the Keck telescope and available through STARLINK. In this way a multiparameter correction model is generated to produce a net of corrections to the drives for the real telescope to ensure perfect tracking and pointing. The net provides precise corrections which are interpolated in real time at a frequency of 20 Hz.

Overlying the model is the Starlink COCO programme which combines the supplied coordinates and epoch with the time supplied by the observatory clock and the location of the observatory to generate the required pointing flow of instructions to image a target object in the centre of the field and track it. The supplied co-ordinates are in terms of right ascension (RA), declination (Dec.) and epoch. Fig. 3 shows the pointing flow taken from Wallace (1987). The pointing flow converts to a standard frame of reference from the given epoch and then calculates the effects of the annual aberration, light deflection and precession and nutation of the Earth. The Earth's rotation and diurnal aberrations are included along with refraction of the atmosphere.

The above calculations are all performed on-line in the telescope drive computer. The system simulates a virtual telescope in which the inputs are the tracking coordinates. The software and the computer obscure the defects of the real telescope by the use of the model. The output of the drive system includes a measure of the difference between the pointing of the real telescope and the requested pointing position.

The controller includes a macro-scheduler which produces an outline schedule for each night of observing. This is then updated by a micro-advisory scheduler which takes into account the observing conditions and the progress of the night's observing programme.

The telescope is now operational (September 1994) and can be accessed using the World Wide Web on the Internet with URL **http://www.eia.brad.ac.uk/rti/**. The system can be operated both remotely and robotically. In the robotic mode it produces comparative photometry on CCD fields for submitted requests from accredited beta test observers. World Wide Web users accessing the site obtain a menu which includes information including an on-line user guide for the telescope with details of hardware and software, an astronomy tutorial package and daily telescope weather reports. Job submission and observation results are password protected. On entering the correct username and password a user can look at their job's progress in the operations database.

The CCD system has a field of about 13 by 19 arc minutes under normal operation and a 12-position filter wheel with Johnson BVRI and Z filters currently installed. The first iteration of the pointing system produced sub arc minute pointing and tracking within the seeing disc of stellar objects at this site. This is adequate for current operations. It is expected to achieve sub 5 arc second pointing and sub arc second tracking as envisaged in the development programme. The current precision of the control system is more than adequate for bright objects. The programme of minimising the pointing accuracy will assist in field confirmation and will enable the system to work confidently in relatively crowded fields. The expected high precision of the system will also support the evolutionary development of pattern recognition software.

References:

Baruch, J., 1992 Robotic Telescopes for Photometry, in: *Variable Star Research—an International Perspective*, eds. Percy, J.R., Mattei, J.A. and Sterken, C., Cambridge University Press, p 109.

Wallace, P.A., 1987, Pointing and Tracking Algorithms for the Keck 10-metre Telescope, in: *Instrumentation for Ground-Based Optical Astronomy; Present and Future*, the Ninth Santa Cruz Summer Workshop in Astronomy and Astrophysics, Springer-Verlag.

A project for an automated telescope in Argentina

J.R. Garcia, S.A. Dominguez and J. Campos

Instituto Copernico, Buenos Aires, Argentina.

Abstract

As we have discussed at the previous Boston Colloquium on Robotic Observatories, a possible project was to place an APT in Argentina. Now, the possible project becomes a real one, and we have started construction. We also describe here the mean characteristics of the site where the APT will be installed and the primary scientific program to be developed by the APT.

1. The telescope

The telescope, as we all know, has two fundamental parts: the optics and the mechanics. Let us begin our description with the first of these.

The configuration adopted for this telescope is a classic Cassegrain, with a primary mirror with a diameter of 400 mm (16"), of 2 metres focal length, which gives a short focal ratio of 5. The secondary mirror has a magnifying factor of 4 ($\gamma=4$), then, the total focal ratio is 20. The figure of the optical surface is made by our own group, but due to the unavailability of a flat and free of tension blank of glass, even Zerodur or Pyrex, we had to put our efforts into making the blank by our own technology. If we had had to buy the blank in the USA or Europe, we should have waited a long time and the expenditure would have surpassed our budget.

In this way, we designed a disc of crystal with a diameter of 400 mm, with a honeycomb structure of 50 mm height. The disc is composed of a compact optical surface of 15 mm thickness, lying over seven hexagonal cells, delimited by ribs 16 mm thick. In Figs. 1 and 2 is shown the plan of the mirror. With this configuration we achieve a saving of 45% in the weight of the mirror, compared to a conventional mirror. For this reason, the disc is more reliable and manageable during the figuring and polishing processes, and we also save energy and time during the disc fusion.

When we considered the design of the mechanical parts of the telescope, we took into account four basic requirements: economy, versatility for automation, hardness and stability in the structure. The economy was achieved by adopting a modular structural configuration, nearly completely built of steel tubes of rectangular profile. To achieve the other requirements, it was necessary to make a detailed study of the problem, in order to select the final configuration to be used. Finally, we adopted a horseshoe mount that completely meets all four requirements. The mounting design has four fundamental modules, that we can state as follows: (1) base; (2) horseshoe structural polar axis; (3) structural tube; (4) mechanical driving devices.

The triangle-shaped base is composed of arc-welded steel tubes of rectangular section, as shown in Figs. 3 and 4. It acts as an orientation and support device. The orientation is

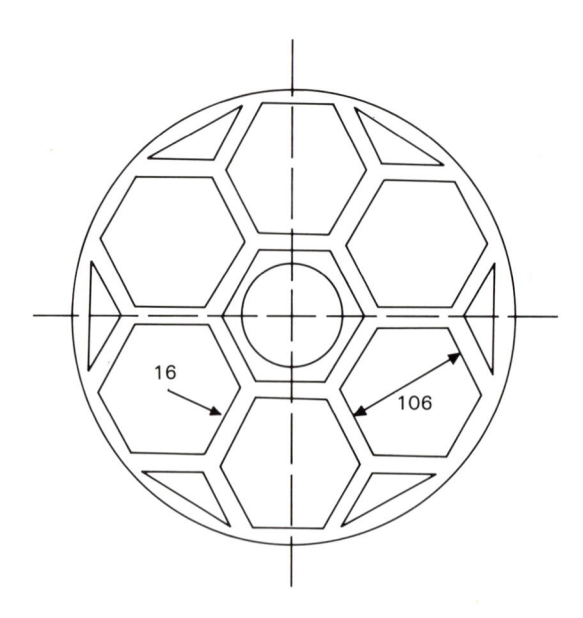

Fig. 1: Rear view of the blank for the primary mirror.

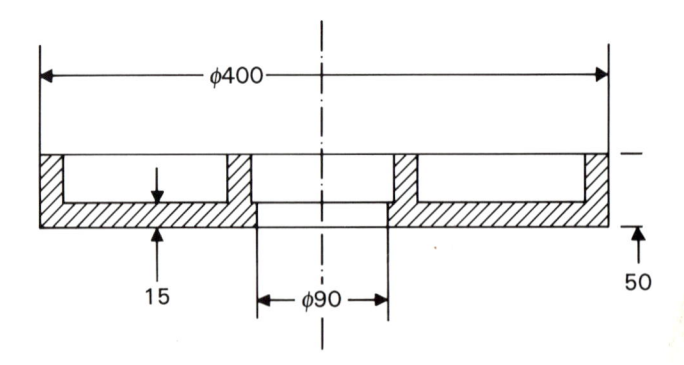

Fig. 2: Cross-sectional view of the blank for the primary mirror.

achieved by means of three indicators that act over the base. One, at the rear, enables fixation of polar height, and two, at the sides and external to the base, enable the fixation of the azimuth of the celestial pole. Once oriented, the instrument is fixed completely by means of a nut and counter-nut system. The horseshoe is supported by two ball-bearing shaft wheels that lie over the front of the base (one side of the triangle), and, at the rear (the opposite vertex), there is an axle-bush that supports the extremity of the structural polar axis.

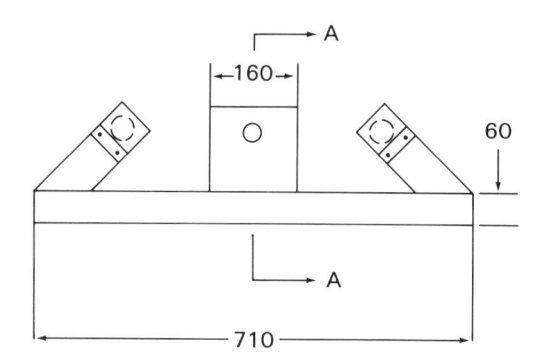

Fig. 3: Front view of the base.

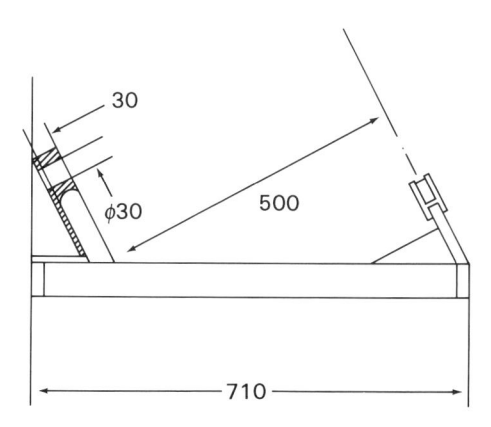

Fig. 4: Cross-sectional (A to A from Fig. 3) view of the base.

The horseshoe—structural polar axis, as shown in Figs. 5 and 6, has two functions of prime importance. First, the horseshoe, constructed from a steel plate of 10 mm thickness, serves as a support of the structural tube; meanwhile its perimeter, turned on a lathe and calibrated, slices over the ball-bearing shaft wheels of the base. The hour motor acts over it, transferring the hour angle tracking to the tube. The structural axis is formed by three steel tubes of rectangular section, placed at 120 degrees one from the other, closed at the rear by a 30-mm compact axis, where the rear axle-bush of the base is located. The coupling between the structural tube and the horseshoe is achieved by two grooved axle-bushes, easy to dismount, where the gudgeon pins of the structural tube are located, enabling the declination movement.

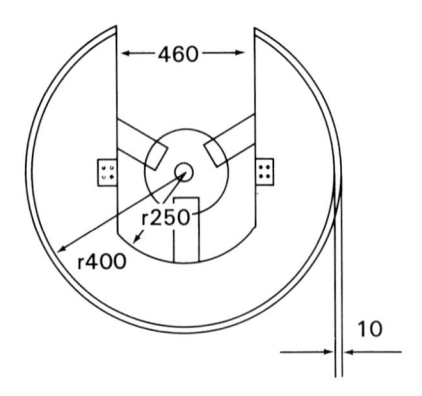

Fig. 5: Front view of the horseshoe–polar axis set.

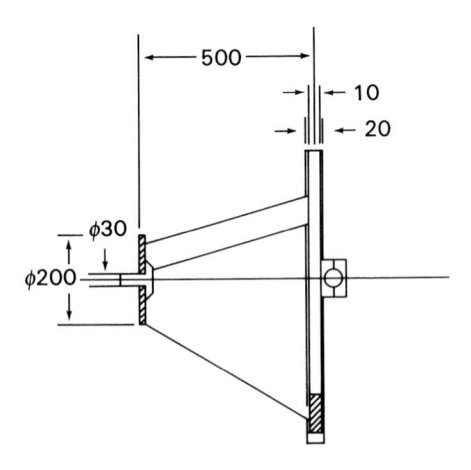

Fig. 6: Cross-sectional view of the horseshoe–polar axis set.

The structural tube, as shown in Fig. 7, maintains the optical elements (primary and secondary mirrors) fixed in their position, and also enables the declination movement. In order to fulfil these tasks, the structural tube comprises two joined modules, lower and upper. The lower module is composed of the structural cell in which is located the primary mirror. The mirror lies in a cell over a floating system with twelve fulcrums, in order to save the primary mirror from deformation. Together with the cell lies the turned horseshoe which we have previously explained. The upper module is formed by a set of ribs that are designed for holding and setting in position the secondary mirror, as well as its focus and collimation systems. Those two modules are coupled by means of a screw set, in order to form a single structure.

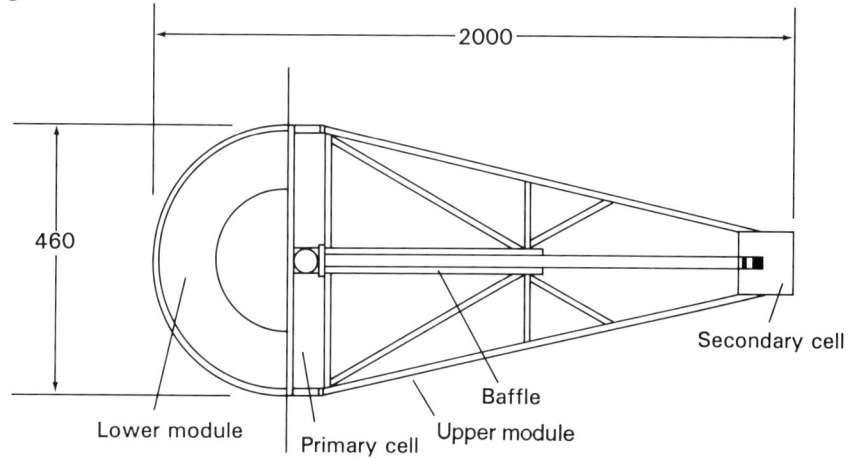

Fig. 7: Lateral view of the structural tube.

The mechanical driving devices are those that give the different movements to the instrument. These movements are achieved by means of different types of motor according to the type of movement to be performed. However, for transmission, we use solely friction wheels, because a gear mechanism is liable to have problems with backlash that are difficult to overcome.

The distribution of the drive systems is as follows: (a) for the hour angle movement, there are two independent motors: one stepper motor for pointing and one synchronic motor for tracking; (b) for the declination movement, there is only one stepper motor for pointing; (c) another stepper motor has been placed over the tube for the optical focusing system.

2. The site

The best site for placing an automatic telescope in Argentina is El Leoncito, San Juan, where two observatories are now in operation. The site is very favourable because of the good atmospheric conditions, and its general characteristics. Its most important natural resource is the atmosphere. Dry and stable upper atmospheric conditions produce stunning clarity in the night sky. The complete absence of artificial light (the site is an astronomical reservation protected from light pollution) makes it most favourable for photometry and imaging. The skies maintain their clarity even if some clouds are present on the horizon, principally due to the region being a dry desert, located between the Andes and a group of Cenozoic mountains called the Precordillera, in the western side of the latter. Dust pollution is not a problem during winter due to the snow. It snows very often, during the day and occasionally during the night. In contrast, the summer is dry and dust pollution increases, giving rise to the poorest seeing of the year. Also, the skies remain more cloudy and the image quality is not so good.

The average percentage of observing nights for the last four years indicates 40% of nights completely clear, and 25% of nights partially cloudy. This means that if we try to undertake photometry with a monochannel photocathode detector, we must consider only the percentage clear sky as useful. However, if we use a field detector, such as a CCD, and try to do only differential photometry, the partially cloudy nights can be considered completely useful. The geographical co-ordinates for the site are: longitude = 04h 37m 20s E; latitude = $-31°$ 48', 08"; altitude = 2411 m.

3. The program

Asteroseismology is a powerful tool which allows the interior structure of stars to be examined in a way analogous to the use of seismology of the Earth. The δ Scuti stars and, particularly, the roAp stars (rapidly oscillating Ap stars) are excellent subjects for asteroseismology studies. The most complete reviews about these stars are in Breger (1979), Kurtz (1990) and Matthews (1991). The H—R diagram (Fig. 8) is constructed from the Garcia et al. (1988) catalogue for this type of pulsating star.

The asteroseismologist should see the δ Scuti stars as a gold mine: many of them are naked-eye stars and very easy to observe. They are also common, many pulsate in non-radial modes and some pulsate in many non-radial modes simultaneously. The latter characteristic is most important since it is the variety of spherical-harmonic modes which gives us the ability to resolve the interior structure of a pulsating star, as emphasised by Kurtz (1988). The oscillations in the roAp stars are usually studied using high-speed photometry through a Johnson B filter.

The basic goal of such time-series photometry is to extract from the light curve of the star the component frequencies of the oscillations. A coherent oscillation gives rise to a peak

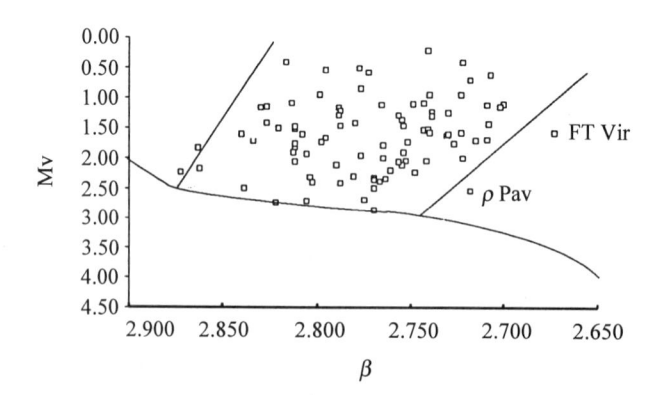

Variable stars in the instability strip near the ZAMS

Fig. 8: H—R diagram for δ Scuti stars. The curve is the ZAMS and the two straight lines represent the boundaries of the instability strip.

in the Fourier transform whose height is proportional to the amplitude of the oscillation. The width (resolution) of such a peak is roughly proportional to the inverse of the length of the light curve. It is possible to improve the resolution by combining data from successive nights, but then the day-time gaps in the data give rise to ambiguous peaks, or aliases, in the amplitude spectrum. Aliasing arises because of cycle count ambiguities when the data are interrupted periodically, as is the case when observations are acquired over several nights from a single site. Since the alias peaks confuse the analysis, the only way to reduce the aliasing problem is to minimise the day-time interruptions in the light curve, hence the need for contemporaneous multi-site observations, as emphasised by Martinez and Kurtz (1991).

Another reason for acquiring contemporaneous multi-site observations of roAp stars is that the oscillation spectra are not stable in many roAp stars. There is evidence of transient oscillation modes with growth/decay times of the order of a day. There is also a possibility of phase jitter in at least one well-studied roAp star. If we are following the temporal behaviour of a changing amplitude spectrum, continuous monitoring of the star is required.

It is important to appreciate that the detection and study of roAp star pulsations demand the most precise ground-based high-speed photometric measurements possible. Although most roAp stars are so bright that photon statistics are not the major source of noise, the amplitudes of oscillations are very low—less than 1 millimagnitude in many cases. In order to produce usable roAp photometry, it is imperative that observers overcome some common sources of error.

The requirements for an adequate observational programme which fulfil the complete set of characteristics of these stars are as follows:

(i) The best time resolution: the magnitude varies during the integrations.
(ii) Differential photometry: the star colours also vary during integration.
(iii) Excellent atmospheric conditions and a careful measure of the sky background: these stars have the extremely small amplitudes.
(iv) Very long monitoring times: these stars are multiperiodic, and they vary their periods from cycle to cycle.

Therefore, the continuous monitoring of these stars is a good challenge for a network of observatories around the world. But the best choice is a GNAT, because this completely covers the requirements. The increasing accuracy of observations from robotic observatories encourages the application of these techniques to the study of the δ Scuti stars. By now, precision better than 1 millimagnitude is achieved by differential photometry, as discussed in this volume and in our recent meetings on automated telescopes. This precision is high enough for performing stellar seismology.

4. Conclusions

In terms of the telescope under construction, we wish to conclude that within a year and a half we will be beginning with the installation at the site. Regarding the site, we wish to encourage projects in order to share our site with other automated telescopes from any institution interested in southern sky objects. Regarding our observational program, we believe that the study and the continuous monitoring of δ Scuti stars may be a perfect challenge for a global network such as the GNAT.

References:

Breger, M., 1979, *Pubs. Astr. Soc. Pacific*, **91**, 5.
Garcia, J.R., Cebral, R., DiGiorgio, F., Romano P., Scoccimarro, E.R., Wahnon, P. and Zimmermann, M., 1988, *Bull. Inform. CDS*, **34**, 67.
Kurtz, D.W., 1988, in: *Multimode Stellar Pulsations*, ed. G. Kovacs et al., Konkoly Obs., Kultura, Budapest, p. 95.
Kurtz, D.W., 1990, *Ann. Rev. of Astron. Astrophys.* **28**, 607.
Martinez, P. and Kurtz, D.W., 1991, *IBVS, No. 3634,* 1.
Matthews, J.M., 1991, *Pubs. Astr. Soc. Pacific*, **103**, 5.

The Fastnet Observatory CCD APT system

E. Ansbro and H. van Bellingen

*Fastnet Observatory, c/o Schull Planetarium,
Colla Road, Schull, Co. Cork, Ireland.*

Abstract

We describe the development of an automatic observatory to be sited at Fastnet in the southwest of Ireland. The telescope is on 11"/16" Super Schmidt camera and will be equipped with a CCD detector.

1. Outline of the project

The telescope at the centre of this project is a 11"/16" f/0.7 Super Schmidt camera. The first stage of development will be to modify the photographic instrument to take a CCD detector. Since a very wide field of view is typical for a Super Schmidt, the instrument will retain a wide field of view even with a CCD detector. A 1000-pixel CCD with 12-micron pixel pitch will deliver a field of view of nearly $4°$ square.

Once the telescope is operational with a CCD detector, an extensive building and testing programme will be implemented to complete the observatory and make the camera telescope an automatic instrument. Since Irish weather conditions do not give too many absolutely clear nights but a reasonable amount of partially cloudy ones, an effective weather station needs to be developed that takes advantage of a maximum amount of clear hours without putting the equipment at risk.

The final phase of the building and testing programme is the development towards remote operation. Once this is successful, the telescope will be moved to a remote site. The authors believe that automatic telescopes will make an increasing contribution to the information pool of observations. To make them more effective, networking has to be kept in mind for the future. For that reason, the approach to the software will be of an open system that, once operational, can easily be connected and networked with other automatic telescopes.

2. The 11"/16" Schmidt camera

By definition, this is a small telescope. This makes it very suitable for automatic operation. The telescope is fork-mounted equatorially. It stands about 4 feet high. As a photographic instrument, the telescope produces images down to 15th magnitude.

2.1 PRINCIPLES OF THE SCHMIDT CAMERA

The main elements of a Schmidt telescope are the spherical mirror, a diaphragm which limits the incident beam and a corrector lens that corrects for spherical aberrations (see Fig. 1). This design gives a high-quality, wide-field instrument. The term "Super Schmidt" points to the very short focal length of the telescope and consequently a very dense field of view. By introducing a CCD in the curved focal plane only a small fraction of the total spherical mirror field of view is used. However, relative to other telescope—CCD combinations this design still offers a wide field of view. The maximum size of the CCD (a flat detector) is limited by the curvature of the focal plane.

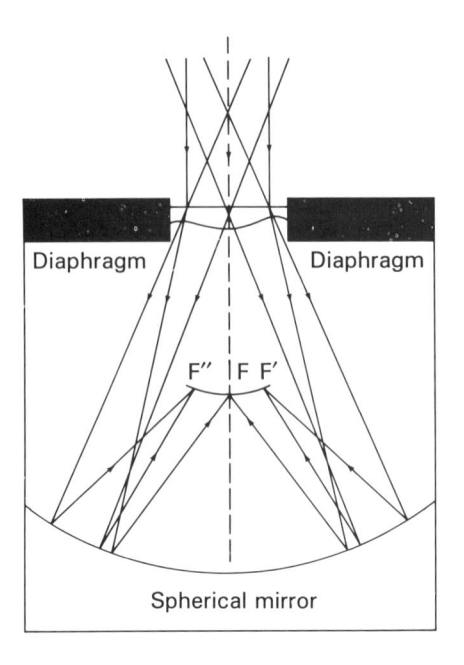

Fig. 1: Schmidt optics.

2.2 THE 11"/16" CAMERA

The Super Schmidt has a 16-inch spherical mirror, an 11-inch corrector lens and a focal ratio of f/0.7.

The RA and Dec. drive use a worm and gear system together with sprockets and belts and a stepper motor and produce a step size of 0.4" in R.A. and 0.6" in Dec. Slewing speed is about $1°$ per second. The drives are controlled by computer, but without the need to programme in assembler. The electronics are made IBM-compatible.

The telescope system is designed with one main computer in mind to control the telescope drive and the CCD camera and smaller processor boards that accomplish various control functions and interface when necessary with the main computer (see Fig. 2).

As data collection with a CCD involves large memory space if stored unprocessed, it is planned to have real-time data processing that will reduce the storage need to a minimum. The proposed observing projects at this moment do not involve storage of images, rather data on individual objects (i.e. patrol work involving asteroids and comets, monitoring of large star clusters, wide-field photometry etc.).

2.3 CHARACTERISTICS OF THE TELESCOPE—CCD COMBINATION

lines as a photographic Schmidt camera's, i.e. patrol work (asteroids, comets, galaxies), monitoring projects (large star clusters and star fields), wide-field photometry.

The limiting factor at this moment is not so much the limiting magnitude but the pixel pitch of the CCD. The field density is about 1" per micron. The image size (which is optics limited) is 25 micron (or 25"). A workable pixel pitch is around 10 micron. A series of tests will be done to determine the astrometric accuracy of this telescope-CCD combination. An acceptable accuracy would be a bonus to patrol programmes.

The array size of the CCD needs to be compatible with the main control computer. During test phase 1, a small CCD in combination with an IBM XT will be used.

3. Towards an automatic observatory

3.1 TOWARDS AUTOMATED OBSERVING

To facilitate the development of the telescope towards automated and finally remote operation, the observatory will be placed on a test site for a period of about 2 years. As this site is near to our workshop, it will make life easier in the early stages. As soon as the CCD-telescope combination is operational, initial elementary automatic observing programmes will be programmed. That allows the system to be monitored and further developed while already observing.

3.2 THE OBSERVATORY SITE AND BUILDING

The final site of the observatory will be near a 500-m high mountain top. This remote place will offer the best skies of southwest Ireland. Since the climate is mild, humidity can be high and appropriate heating systems need to be installed to prevent condensation on optical components.

General weather conditions should see the observatory operating on average 1 day out of 2. This precludes sophisticated weather monitoring in order to have the observatory operational during clear spells.

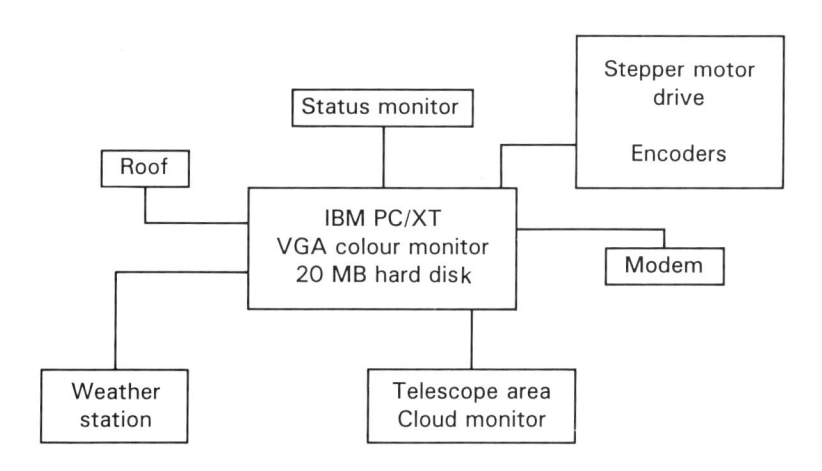

Fig. 2: Telescope control systems etc.

Because of the small size of the telescope the observatory building can also be kept small. This is an advantage as general weather conditions in winter include storm force winds up to 200 km/h. The building will be developed and tested at the test site. When all parts live up to expectations, it will be disassembled and rebuilt at the final site. We hope that this way the starting up problems will be reduced to a minimum.

3.3 TOWARDS REMOTE AND AUTOMATIC OPERATION

Once observational programmes with satisfactory low data output are all programmed and the complete system responds well, our efforts will go towards finalising the software

4. Software and observing projects

It is impossible to plan for all the different projects that an instrument can or will perform. However, it is possible to design software that will allow an easy change in a certain area. That is why the general approach to the software is one of a modular structure where individual elements can easily be changed for a more advanced or different version without upsetting the rest of the program. The programming language is Pascal and the use of machine language will be avoided.

The fine-tuning of software and hardware is very important. This is of course dependent on the actual computer in use. Thus programs must allow for easy change of approach if a faster computer is introduced. It is of no use to take an image in 30 s and process it in 2 min as this decreases the efficiency of the telescope. Since our aim is to have real-time processing, this might lead to a different approach for a specific observing project. Patrol work will incur the difficulty of recognising fast enough known objects or objects of fixed position.

5. Areas for co-operation

Since the Fastnet Observatory consists of a small group of dynamic people, it is open for co-operation in several fields. Specifically, we would like to communicate with others in the field of electronic hardware and CCD detector problems. Any ideas in the field of software design and development are welcome since it is our aim to allow outside use of the instrument.

It is evident that we will encounter all sorts of problems in the development of the necessary hardware and software. Since we take an open approach ourselves we are delighted to hear from the experience of others. For a detailed technical description of the project, please do not hesitate to contact us at Fastnet Observatory.

PART II

OBSERVATIONAL RESULTS

The quest for precision robotic photometry

G.W. Henry[1] and D.S. Hall[2]

[1]*Center of Excellence in Information Systems,*
Tennessee State University, Nashville, TN 37203, USA.
[2]*Dyer Observatory, Vanderbilt University, Nashville, TN 37235, USA.*

Abstract

We discuss limitations of early automatic photoelectric telescopes and document the improvement in precision of data from the Vanderbilt/Tennessee State 16-inch robotic telescope over the past few years. Through careful temperature control of the photomultiplier and filters, accurate centering and tracking of stars in the diaphragm of the photometer, and telescope control software that allows nightly determination of extinction coefficients, we have succeeded in obtaining high-quality photometric data that are limited only by photon and scintillation noise.

1. Introduction

It is now just over a decade since the night in Phoenix, Arizona, when Lou Boyd's 10-inch automatic photoelectric telescope first observed a list of chromospherically active stars and revolutionized the way we do astronomy (Boyd et al. 1984). From 12—13 October 1983 onward, it would no longer be necessary to spend large portions of our time with manual telescopes on remote mountaintops gathering brief glimpses of the behaviour of these enigmatic objects. Instead, Boyd's Phoenix-10 and later the Vanderbilt/Tennessee State 16-inch robotic telescope on Mt. Hopkins (Hall 1989) would not only provide year-round, inexpensive photometric coverage of our stars but also free us to spend time interpreting data instead of gathering it. Hall and Henry (1993) document 47 new variable stars discovered with these two robotic telescopes and provide a partial listing of publications relating to a list of important dates in their history. Between the fall of 1987 and the summer of 1992, the 16-inch made 36,000 good group observations (as described below) of chromospherically active stars with an operational cost of approximately $1.50 per group. By comparison, human observers with conventional manual telescopes would cost about $20 per group observation. Currently, 140 program stars are on the menu of the 16-inch telescope, and automatic routines for observing, data reduction, quality control, and archiving of the 16-inch data to optical disk require literally only seconds per day to verify that data acquisition is proceeding normally.

Besides the advantages of unattended observing with a dedicated telescope at a prime photometric site and the resulting deluge of data coming to us, we have succeeded during the past two years in greatly improving the precision of these data as well. Although many problems have been encountered along the way (see Hall and Henry 1993 for a detailed list), robotic telescopes have become extremely reliable instruments capable of high-precision photometry. As we will show below, improvements to the 16-inch telescope, photometer, and observing methods have resulted in precision measurements limited only by scintillation and photon noise.

2. Limitations of early robotic telescopes

The first three automatic telescopes operated by Fairborn Observatory were the 10-inch and 16-inch telescopes mentioned above along with an additional 10-inch belonging to Fairborn. By the end of 1987 they were all in operation at the new Fairborn Observatory site provided by the Smithsonian Astrophysical Observatory on Mt. Hopkins in southern Arizona (Genet and Hayes 1989). All three were operated identically by a control program described by Boyd et al. (1984) which selected the next "group" of stars to be observed based on a simple but effective "first to set in the west" rule. Each group observation consisted of eleven moves between a check star (K), sky (S), comparison star (C), and variable star (V) in the following fixed sequence: K,S,C,V,C,V,C,V,C,S,K. Each star was centered in the diaphragm by taking counts with the photometer in four overlapping regions of the sky centered around the presumed position of the star. The counts in these regions were compared, the direction to the true star position derived, and the telescope moved to that position. This process was repeated until the count rates in all four regions were the same, indicating that the telescope had properly centered the star. The accuracy of this centering process was probably about one-quarter of the diaphragm diameter. The fixed sequence of measurements within a group and the group selection algorithm prohibited the observation of standard stars during the night so the differential observations were reduced with mean extinction and transformation coefficients determined on occasional nights dedicated to standard star observations.

Besides the limitations imposed by the telescope control software, there were further limitations to the precision of the data resulting from telescope hardware. The two primary problems with the Vanderbilt/Tennessee State 16-inch were the worm gear drive systems and the lack of temperature control of the photometer. The 16-inch telescope axes were originally driven by small anti-backlash worm gears (Henry et al. 1991). The heavy demand placed on these drives in the automatic observing mode resulted in continual wear of the gears. This made it increasingly difficult for the preloaded drives to resist motion due to wind loading or slight imperfections in telescope balance. As will be seen below, this resulted in a gradual degradation in the precision of the measurements during the three years of operation with the worm gear drives. The photometer of the 16-inch featured a thermoelectrically cooled EMI 9924B photomultiplier tube. However, the temperature was not controlled tightly enough to prevent temperature variations of the tube with large changes of ambient temperature, nor was any attempt made to stabilize the temperature of the filters. Therefore, night-to-night and seasonal changes in the transformation coefficients were significant sources of error in our differential magnitudes reduced with mean coefficients.

In spite of their limitations, however, these telescopes produced an enormous quantity of good photometry (see Hall and Henry 1992 for selected references). The external errors (as defined below) for the Phoenix-10 and the Fairborn-10 were documented in Young et al. (1991) to be in the range from 0.010 mag to 0.015 mag. Henry et al. (1991) showed that the external errors for the Vanderbilt/Tennessee State 16-inch with the original worm gear drives and old photometer were in the same range.

3. Upgrades to the Vanderbilt/Tennessee State 16-inch Telescope

Three significant upgrades have been made to the 16-inch during the past two years in the effort to improve the precision of the data as well as the efficiency of the telescope.

The first involved the installation of new stepper motor driven sprocket and (Berg) belt drive systems to both axes in the summer of 1990. These new drive systems are similar to ones described by Genet and Hayes (1989) for the newest generation of robotic telescopes. They provide better resistance to wind loading, are free of backlash, allow much faster slewing of the telescope, and eliminate the wear associated with worm drives. Centering and tracking of stars were both improved with the installation of these new drives. The increased speed improved the observing efficiency by 10% to 15%.

The second improvement, also accomplished in the summer of 1990, consisted of a new telescope control computer and a high-speed modem link from Arizona to Tennessee State. The new computer allowed execution of the new telescope control software package developed by Fairborn Observatory called the Automatic Telescope Instruction Set or ATIS (Genet and Hayes 1989). With ATIS, we gained full control of the observing program and could modify our differential observing sequences as needed, interperse standard star observations throughout the night, run engineering tests, do all-sky photometry, monitor a particular object for several hours, or do any combination of these within a given night. The modem link gave us daily access to the data as well as freedom to adjust the observing program whenever desirable.

The most recent upgrade was the installation of a new precision photometer designed by Lou Boyd of Fairborn Observatory during the summer and fall of 1991. This new photometer uses a CCD camera for finding stars and quickly centering them in the focal plane diaphragm. Centering times are improved from around 10 to 15 seconds with the old centering routine to about two seconds with the CCD camera, and centering accuracy is now around two arcseconds. Once a star is properly centered in the diaphragm, a rotating mirror admits light to the photomultiplier to make the observation. Accurate temperature control is provided by a combination of thermoelectric coolers and circulating glycol that stabilize the photomultiplier and filters to within a few tenths of a degree. Heat can be added or removed depending on the ambient temperature to insure stability. Filtered and dried air is constantly circulated inside the sealed enclosure of the photometer to control dust and humidity. Two rotating filter wheels and one rotating diaphragm wheel provide a large selection of colour and neutral density filters and diaphragm sizes. Routine observations with this new photometer began in early 1992, and approximately one-third more stars could be added to the observing program due to the greater efficiency of CCD centering. Comparison with our old observing notebooks from Dyer Observatory and the Kitt Peak No. 4 16-inch indicate the robotic telescope is now gathering data at a rate several times faster than our best manual efforts.

4. Improvements to the precision of the 16-inch data

Since each group observation consists of three differential measures between a variable star and its comparison, we can use the standard error of a group mean (SEM) as a measure of the internal precision of our data. The SEM is calculated from the standard deviation of the three differential measures as in Hall, Kirkpatrick, and Seufert (1986). During the course of the automatic reductions on each night's data, the average standard error of all the group means for the night is computed and tabulated with results from previous nights. The telescope observes whenever it can find stars, so these nightly SEMs are useful for flagging nights of poor photometric quality. Cloud filtering of poor data is done in the final archiving process when group means with a standard error greater than 0.01

mag are rejected. Because the stars on our program can be assumed to be constant over the six minutes it takes to complete a group observation, these SEMs are an effective measure of the nightly internal precision of the 16-inch observations.

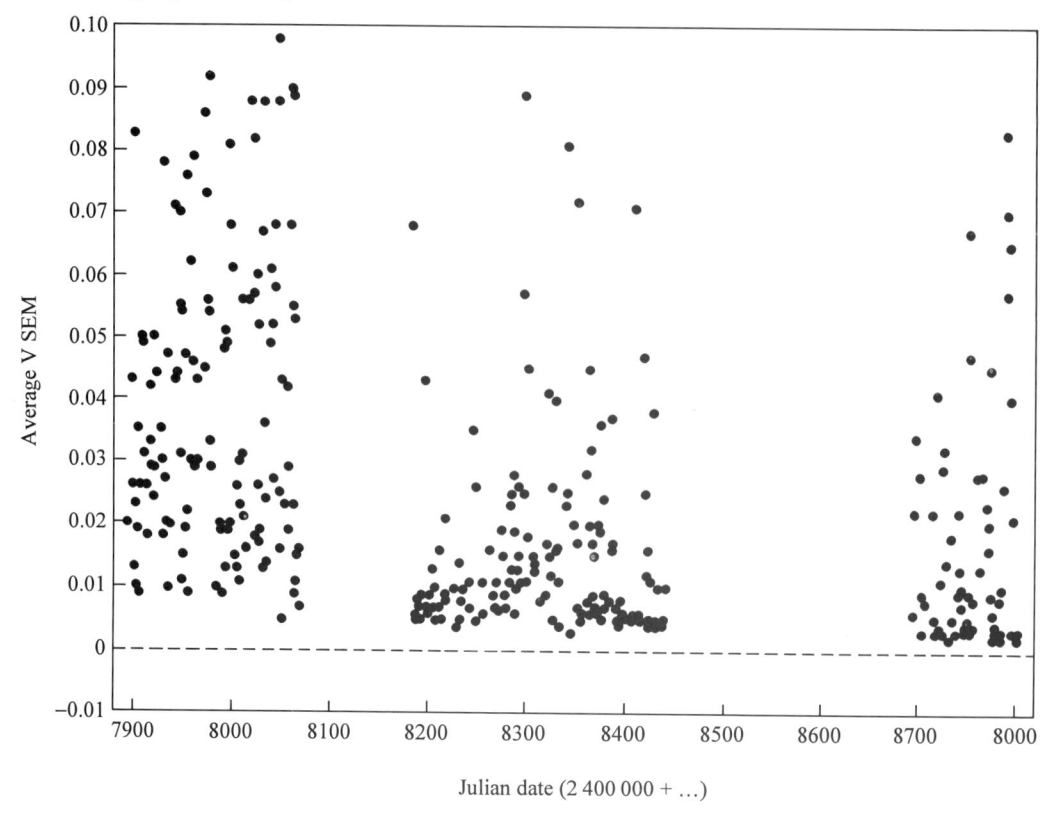

Fig. 1: The nightly average standard error of the group means (SEMs) in V for the Vanderbilt/ Tennessee State 16-inch robotic telescope. The bottom of the distribution of SEMs represents the internal precision of the data on the best nights and shows the improvement when a new drive system was installed (during the first gap) and when the precision photometer began operation (after the second gap).

Fig. 1 plots the nightly SEMs for the V filter from the beginning of 1990 through mid-1992. The standard errors of all group means are included in these nightly averages whether or not any individual observation will later pass the cloud filter. The bottom of the distribution of SEMs, therefore, represents the internal precision of the system on the best nights. The first gap in the record occurs when the original worm drives were replaced with the new belt drives. The second gap occurs when the telescope was fitted with the precision photometer. Since lack of temperature control of the old photometer would not be expected to affect the internal precision of a group measurement made over only six minutes, this plot shows clearly the gain in precision derived from improvements to the centering process. Prior to the drives upgrade, there were very few nights when the average internal error was better than about 0.01 mag, although the portion of the record shown here before the first gap was obtained with the original worm gear drives in their worst condition just before replacement. After the drives upgrade, the internal precision improved to about 0.005 mag on good nights. While the original crude method of centering was still in use, the improved drives eliminated backlash and were better able to maintain the centering during integrations. With the advent of the new photometer and the use of a CCD camera for rapid and precise centering, the internal precision improved further to about 0.003 mag and occasionally to 0.002 mag.

The external precision of the 16-inch observations can be estimated from the standard deviation of the individual nightly group means of presumed constant stars from their long-term seasonal mean. These external errors provide a measure of the effectiveness of the nightly extinction corrections and the long-term stability of the transformations to the standard system. Two constant star pairs were placed on the observing menu of the 16-inch to monitor these external errors. The first pair of stars (27 and 28 LMi) has been on the menu since the beginning of operations and consists of A6V and KOIII stars with a difference in colour index of 1.0 mag. The second pair (14 and 18 Boo) consists of F6IV and F5IV stars with colour indices differing by only 0.1 mag and was placed on the menu after the drives upgrade.

Table 1—Summary of precision (in mag) for 16-inch robotic telescope

Observing Season	σ(in)	σ(ext) (27/28 LMi)	σ(ext) (14/18 Boo)	Comments
1987—88	—	0.011	—	original worm drive
1988—89	—	0.012	—	original drive degrading
1989—90	0.010	0.015	—	original drive at worst
1990—91	0.005	0.009	0.007	belt drive upgrade
1991—92	0.003	0.003	0.003	precision photometer

Table 1 traces the history of the precision of the 16-inch telescope in the B filter. The external errors of the red-blue pair show a steady deterioration during the first three observing seasons due to wear of the original worm drives that affected the internal errors of all measurements. An immediate improvement is evident after the drives upgrade and another very significant improvement occurred after the precision photometer was installed. The external error of the first season of observation of the F-star pair is 0.002 mag smaller than the corresponding error of the red-blue pair since the large colour difference of the red-blue pair makes the transformed magnitude difference more sensitive to changes in colour response of the photometer due to night-to-night and month-to-month temperature changes. Once the precision photometer was installed and temperature changes of the detector and filters were eliminated and nightly extinction coefficients were applied, both pairs of stars had external errors of only 0.003 mag. It should be noted that the 0.003 mag external error is reached only by rejecting observations on poor or marginal nights when good nightly extinction coefficients could not be determined, even though the individual group means may have passed the 0.01 mag cloud filter. When this is done, external errors do not appear to be significantly larger than the internal errors.

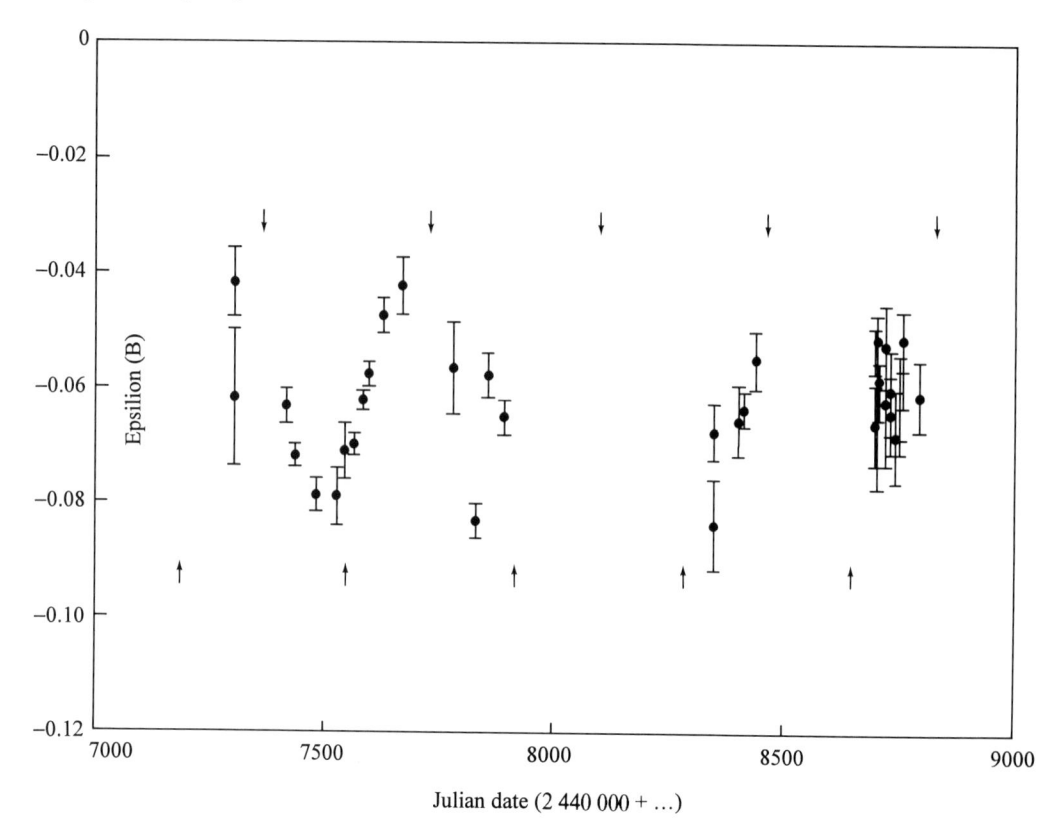

Fig. 2: The transformation coefficient in B for the 16-inch telescope over a period of nearly 5 years. Downward-pointing arrows mark the approximate time of mid-summer; upward-pointing arrows mark mid-winter. Seasonal variations of the transformation coefficient are quite evident prior to JD 2448500 when the temperature-controlled precision photometer was installed. After that time, the coefficient stabilized at a value intermediate between the summer and winter extremes.

Fig. 2 demonstrates the effect of changing ambient temperature on the transformation coefficient in B and the stabilization of the coefficient once the temperature-controlled precision photometer went into use. Prior to JD 2448500, transformations were determined with the old photometer on occasional nights dedicated to standard star observations. Seasonal variations in the coefficient are quite evident. In fact, periodogram analysis of these data reproduce the length of the year quite well. After JD 2448500, nightly observations of approximately 50 standard stars (requiring only a few percent of the telescope time) with the precision photometer allow nightly solutions for extinction and transformation. It can be seen in Fig. 2 that the B transformation coefficient has stabilized at a value intermediate between the summer and winter extremes. The nightly mean differential magnitudes between the red-

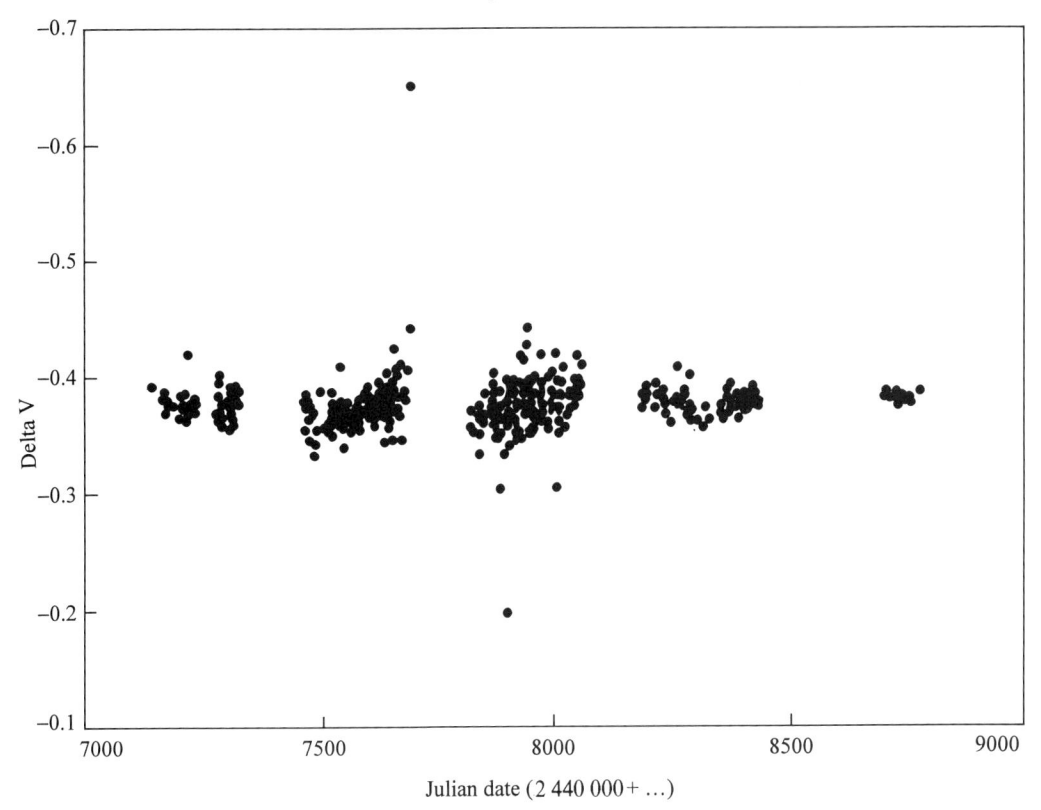

Fig. 3: The scatter in the nightly mean differential magnitudes between the red-blue pair, 27-28 LMi, over five observing seasons measures the external precision of the 16-inch data. The effect of wear in the worm drives during the first three seasons is evident as well as the improvement when the belt drives were installed for the fourth season. The fifth season shows the further gain due to CCD centering, temperature control of the new photometer, and application of nightly extinction coefficients.

blue pair, 27-28 LMi, over the past five observing seasons are shown in Fig. 3. Here the effect of wear in the original worm gear drives is clearly evident in the growing scatter during the first three seasons. The fourth season illustrates the improvement realized with the new belt drive systems. The fifth season shows the further gain in external precision due to CCD centering, temperature control of the new photometer, and application of nightly extinction coefficients.

5. Photon and scintillation noise limits

We can estimate separately what the contributions to the nightly average internal error should be resulting from photon noise and scintillation. The average magnitude of all stars on the observing program of the 16-inch is 6.8 in V resulting in approximately 70,000 counts per second in the V filter. Most group observations are made with 10-second integration times, so the average uncertainty of a group mean due to photon noise should be close to 0.0010 mag. The telescope is programmed to avoid taking observations at an air mass greater than 2.0. The median value of the air mass on a typical night is about 1.35. Young (1974) provides a method of estimating the scintillation noise of an observation as a function of telescope aperture, observatory altitude, integration time, and air mass. At a median air mass of 1.35 and an integration time of 10 s, the 16-inch telescope on Mt. Hopkins should produce a group mean with a predicted uncertainty of about 0.0025 mag due to scintillation noise, after correction for a factor of two error in Young's original formula (Young 1991). On the very best nights on Mt. Hopkins, if we assume that the only sources of error in our group means arise from photon and scintillation noise, then the average standard error of the group means should be around 0.0027 mag. This agrees extremely well with the 0.002 mag to 0.003 mag average standard errors that are actually observed on the best photometric nights. Therefore, on these good nights, the precision of the data is limited only by photon and scintillation noise, and we conclude that the engineering performance of the telescope is no longer limiting the precision of the data. The only way to further increase the precision is simply to spend more time integrating.

Acknowledgements

The 16-inch telescope was obtained with NSF research grant AST 84-14594 to Vanderbilt University. Its continued operation and analysis of data are supported by NASA grant NAG8-111 and NSF grant HRD-9104484 to Tennessee State University.

References:

Boyd, L.J., Genet, R.M. and Hall, D.S., 1984, *IAPPP Comm. No. 15*, 20.
Genet, R.M. and Hayes, D.S., 1989, *Robotic Observatories*, Mesa: AutoScope.
Hall, D.S., 1989, in: *Automatic Small Telescopes*, eds. D.S. Hayes and R.M. Genet, Mesa: Fairborn Press, p. 65.
Hall, D.S. and Henry, G.W., 1993, in: *IAU Coll. No. 136, Stellar Photometry*, eds. C.J. Butler and I. Elliott, Cambridge University Press, p. 205.
Hall, D.S., Kirkpatrick, J.D., and Seufert, E.R., 1986, IAPPP Comm. No. 25, 32.
Henry, G.W., Nagarajan, R. and Busby, M.R., 1991, *IAPPP Comm. No. 45*, 11.
Young, A.T., 1974, in: *Methods of Experimental Physics*, vol. 12A, *Astrophysics, Optical and Infrared*, ed. N., Carleton, New York: Academic Press, p. 101.
Young, A.T., 1991, private communication.
Young, A.T., Genet, R.M., Boyd, L.J., Borucki, W.J., Lockwood, G.W., Henry, G.W., Hall, D.S., Smith, D.P., Baliunas, S.L., Donalue, R. and Epand, D.H., 1991, PASP, **103**, 221.

The law of starspot lifetimes

D.S. Hall[1] and G.W. Henry[2]

[1]Dyer Observatory, Vanderbilt University, Nashville, TN 37235, USA.
[2]Center of Excellence in Information Systems,
Tennessee State University, Nashville, TN, 37203, USA.

Abstract

We consider 112 starspots on 26 different spotted stars, including the sun, and find that spot lifetimes seem to be governed by two laws and that a given spot's lifetime will be lesser of the two values generated by those two laws. In the first law, based on disruption by the shearing of differential rotation, lifetime is a function of the spot's angular radius, the star's differential rotation coefficient, and the star's rotation period. In the second law, lifetime is a function of the spot's angular radius and the star's linear radius. Our sample of observed lifetimes is represented quite well by this two-part law, with an rms deviation less than 0.3 in the log.

1. Introduction

In an earlier paper Hall and Busby (1990) computed starspot lifetimes (t_c) with the assumption that they are disrupted by the shearing of differential rotation and compared them with observed lifetimes (t_o) for 40 spots on 17 different stars. The differential rotation coefficient (k) for each star was estimated from the range of rotation periods exhibited by its spots. The radius of each spot (r_s) was estimated from its maximum wave amplitude (Δm). They found that small spots ($r_s < 20°$) die before they are disrupted ($t_o < t_c$) whereas larger spots ($r_s > 20°$) are, or seem to be, disrupted by differential rotation ($t_o = t_c$).

In this paper we consider a larger number of spots (112) on a larger number of different spotted stars (26), including the sun, and analyze the starspot lifetime question somewhat more rigorously.

The result is that starspot lifetimes seem to be governed by two laws and that a given spot's lifetime will be lesser of the two values generated by those two laws. In the first law the lifetime (t_1) is a function of the spot's angular radius (r_s), the star's differential rotation coefficient (k), and the star's rotation period (P). In the second law the lifetime (t_2) is a function of the spot's angular radius (r_s) and the star's linear radius (R_*). Our sample of observed lifetimes is represented quite well by this two-part law, with an rms deviation less than 0.3 in the log.

2. Input Data

Table 1 lists the 26 stars we considered. The second column is the rotation period, taken as the orbital period for stars rotating synchronously in a binary or (in brackets) as the mean rotation period defined by several spots in the case of single stars or stars rotating asynchronously in a binary. For those known to be SB-1 or single, L = 1.0. The fourth The third column is the fractional light contributed by the potted star, used to transform the wave amplitude observed in composite light into the intrinsic wave amplitude. The fourth column is the range of spot rotation periods observed, as a percent of the rotation period itself. The fifth column is the number of spots observed in establishing this range. The sixth column is the differential rotation coefficient derived from the entries in the preceding two columns, as explained in the next paragraph. The seventh column is the spotted star's radius, in solar units, and the last column is the principal reference used for these entries, where "HB" = Hall and Busby (1990).

Table 1—Input for spotted stars

Star	P (rot.) (days)	L	Δ P/P (%)	n	k	R ($R_\odot$)	Source
λ And	<53.86>	SB-1	7.6	9	0.080	21.3	in prep.
SS Boo	7.606	0.6	0.55	4	0.0068	3.3	HB
SV Cam	0.593	0.95	0.22	4	0.0027	1.2	HB
BM Cam	80.9	SB-1	1.2	6	0.0133	25.	in prep.
FK Com	<2.398>	single	0.5	4	0.0062	10.	HB
CG Cyg	0.631	0.5	0.025	3	0.00035	0.9	HB
V1764 Cyg	40.133	SB-1	0.63	1	0.0126	23.	HB
V1817 Cyg	108.854	0.98	8.5	2	0.17	62.	AJ **100**, 2017
BY Dra	<3.827>	0.75	10.5	14	0.107	0.75	HB
DK Dra	64.44	0.5	4.5	11	0.047	14.	in prep.
EI Eri	1.947	SB-1	2.5	16	0.025	2.7	ApJ **348**, 682
EK Eri	<335.>	single	—	1	[0.2]	10.	HB
ε Eri	<11.15>	single	20.6	5	0.238	0.8	AJ **102**, 1813
σ Gem	19.423	SB-1	5.35	17	0.054	11.1	in prep.
MM Her	7.960	0.42	0.38	2	0.0076	2.8	HB
RT Lac	5.074	0.5	0.42	8	0.0045	4.	HB
HK Lac	24.428	SB-1	1.15	4	0.0142	7.7	HB
V478 Lyr	2.131	0.99	0.95	9	0.010	0.9	AJ **99**, 396
VV Mon	6.051	0.64	0.48	1	0.0096	6.	HB
V1149 Ori	53.58	SB-1	4.0	6	0.044	13.	AJ **102**, 1808
DM UMa	7.492	SB-1	0.2	2	0.0040	4.	HB
HD 8358	0.516	0.5	0.88	3	0.0125	0.9	IBVS 2773
HD 163621	3.304	0.7	2.9	5	0.0335	1.0	in prep.
HD 181943	<385.>	single	—	1	[0.2]	7.	ApJS **74**, 225
HD 191011	<19.22>	single	29.2	8	0.31	16.	AJ **104**, 1936
sun	<25.4>	single	—	—	0.185	1.0	Allen (1973)

The differential coefficient k is defined by the relation

$$P(\phi)/P(\phi=0°) = 1 / (1 - k \sin^2\phi) \qquad (1)$$

commonly used to describe differential rotation in the sun, for which k = 0.18. If spots occur over the entire 90° range of latitude, and previous work shows that they do in heavily spotted stars (Gray 1988, Table 7-1), and if the spots observed on a given star actually sample the entire 90° range or nearly so, then

$$\Delta P/P = k. \tag{2}$$

If only a few spots are observed, they probably sample a restricted range. For example, three spots might be at $\phi = 22°\!.5$, $45°\!.0$, and $67°\!.5$, in which case

$$\Delta P/P = k \cdot f, \tag{3}$$

where, in this n = 3 example,

$$f = \sin^2\phi(\max) - \sin^2\phi(\min) = \sin^2 67°\!.5 - \sin^2 22°\!.5 = 0.71. \tag{4}$$

Values of f for various numbers of spots would be

n	f	n	f	n	f
2	0.500	8	0.940	14	0.978
3	0.707	9	0.951	15	0.981
4	0.809	10	0.960	16	0.983
5	0.866	11	0.966	17	0.985
6	0.901	12	0.971	18	0.986
7	0.924	13	0.975	∞	1.000

based on this scheme.

Because EK Eri and HD 181943 are single stars and only one starspot has been observed on each, we could not determine k explicitly. The value of k appearing in Table 1 is a reasonable estimate, close to the solar value.

Table 2 is a list of the 112 starspots we considered. Column 1 is the star. Column 2 is the maximum radius, in degrees, which the spot attained, derived from the wave's maximum amplitude following the same procedure used by Hall and Busby (1990, section 3.3). Column 3 is the spot's observed lifetime in years. Our procedure for using a spotted star's migration curve to determine spot lifetimes has been explained elsewhere, a good example being our treatment of HD 37824 = V1149 Ori (Hall et al. 1991)

For spots on stars with the reference "HB" in Table 1, we took the values of r_s and t_o as they appeared originally in Hall and Busby (1990, Table 2). In an attempt to have the largest possible sample of starspot lifetimes in this study, we assigned lifetimes to some spots in cases where only its birth or its death was observed. Our procedure was to assume $t_o = 1.5 \Delta t$, where Δt is the time interval during which the spot was observed to live. The number of spots, n in Table 1, used to derive k from $\Delta P/P$ is not in every case equal to the number of spots with lifetimes given in Table 2. This is because in some cases a spot's rotation period could be determined but incomplete phase coverage or inadequate analysis precluded determination of the spot's lifetime. Parameters for the one sunspot in Table 2 are the 0.020 $R_\odot$ radius of the "mean sunspot group" considered by Allen (1973, section 87) and the 6-day life of an "average sunspot group" given by Allen (1973, section 88).

Table 2—Input for starspots

Star	r_s (deg)	$\log t_o$ (yr)
λ And	17.4	0.079
	15.5	0.290
	18.6	0.455
	23.7	0.455
	14.8	0.462
	16.5	0.477
	15.5	0.484
	19.8	0.829
	24.5	0.989
SS Boo	30.0	0.301
	25.0	0.477
SV Cam	32.0	−0.482
BM Cam	13.4	0.299
	15.3	0.476
	14.4	0.683
	14.8	0.688
	18.8	0.717
	22.0	0.800
FK Com	23.5	0.000
CG Cyg	19.7	0.477
	20.9	0.602
	23.5	0.653
V1764 Cyg	15.7	0.903
V1817 Cyg	10.2	0.301
	8.5	0.398
BY Dra	8.4	−1.356
	10.1	−1.337
	9.5	−0.921
	11.8	−0.699
DK Dra	28.8	0.149
	29.0	0.149
	18.7	0.201
	29.1	0.201
	21.4	0.325
	23.1	0.360
	27.1	0.378

Table 2 (continued)

Star	r_s (deg)	$\log t_o$ (yr)
	29.9	0.393
	33.5	0.548
	22.0	0.569
	25.6	0.569
EI Eri	16.5	−0.301
EK Eri	20.8	1.176
ϵ Eri	8.4	−1.086
	8.7	−1.071
	6.3	−0.939
	7.5	−0.724
	9.6	−0.672
σ Gem	8.7	−0.523
	7.8	−0.260
	14.7	−0.187
	15.9	−0.187
	16.5	−0.187
	13.1	−0.071
	12.3	−0.046
	15.2	−0.022
	14.3	0.146
	17.0	0.279
	8.3	0.290
	16.5	0.301
	17.4	0.301
	18.0	0.301
	14.9	0.470
	18.8	0.538
	16.5	0.732
MM Her	22.0	0.301
	29.0	0.301
RT Lac	25.6	0.602
	19.7	0.845
	19.7	0.845
	25.6	0.903
	20.9	0.954
	30.4	0.954
	29.5	1.114
	24.5	1.204

Table 2 (continued)

Star	r_s (deg)	$\log t_o$ (yr)
HK Lac	27.0	0.518
	28.0	0.532
	26.8	0.602
	30.7	0.634
V478 Lyr	11.0	−0.328
	13.6	−0.237
	12.9	−0.066
	15.4	0.017
	12.2	0.041
	14.2	0.061
	16.9	0.292
VV Mon	16.0	0.778
V1149 Ori	22.0	0.352
	16.4	0.538
	19.3	0.556
	28.2	0.653
	25.7	0.756
DM UMa	25.1	0.699
	29.5	0.699
HD 8358	28.3	−0.807
	30.5	−0.550
	27.4	−0.775
	13.4	−0.404
HD 163621	8.9	−0.801
	16.0	−0.625
	16.8	−0.414
	14.0	−0.114
	13.2	0.036
HD 181943	19.8	1.301
HD 191011	8.5	−0.824
	7.3	−0.648
	6.3	−0.569
	8.4	−0.168
	10.7	−0.168
	9.3	−0.143
	10.8	−0.068
	11.0	0.270
Sun	1.8	−1.785

3. The first law

Lifetimes computed with the first law are the disruption times computed originally by Hall and Busby (1990). In general

$$t_1 = t' \, P(\phi=0°) / k, \qquad (5)$$

where t' is multivalued depending on the latitude of the spot, which generally is not known or known only poorly from spot modelling. Hall and Busby (1990, Fig. 5) gave relations for three special cases: the spot touches the pole, the spot lies at $\phi = 45°$, and the spot touches the equator, the last two giving the shortest and longest lifetime extremes, respectively. The $\phi = 45°$ case can be expressed analytically as

$$1/t' = 0.0285 \, r_S \qquad (r_S < 20°) \qquad (6a)$$

or

$$1/t' = 0.0255 \, r_S + 0.06 \qquad (r_S > 20°) \qquad (6b)$$

In the extreme special case where a spot lies at $\phi = 45°$ and has an angular radius of $r_S = 45°$ (in which case it touches both the pole and the equator) and where $k = 1$, the spot will be disrupted in about one stellar rotation.

4. The second law

This law, which says that relatively small spots have lifetimes shorter than would be allowed by the first law, was cast in the functional form

$$\log t_2 = a \, r_S + b \, \log R_* + c. \qquad (7a)$$

To evaluate the coefficients a, b, and c we used least squares to fit the R_* values from Table 1 and the t_o, r_S values from Table 2. Any spot for which $t_1 < t_2$ was excluded from the fit. The result was

$$\log t_2 = \underset{\pm\,0.01}{0.13 \, r_S} + \underset{\pm\,0.06}{0.49 \, \log R_*} - \underset{\pm\,0.15}{1.95}, \qquad (7b)$$

where t_2 is in years, r_S is in degrees, and R_* is in solar radii. Some iteration was required in this process because obviously the coefficients had to be known before the $t_1 < t_2$ exclusion criterion could be applied, but the convergence was swift.

5. Comparing t_o with t_1 and t_2

Fig. 1 is a log plot of t_o versus t_1, with filled circles indicating spots for which $t_1 < t_2$ and pluses indicating spots for which $t_1 > t_2$. In this case it is the filled circles which should follow the line of perfect correlation, and they do nicely, with an rms deviation of only 0.31 in the log. In this figure the pluses should fall below the line of perfect correlation, and virtually all of them do. These are mostly the smaller spots which do not live as long as their disruption times would permit. Note the generic (small) sunspot which lives only 6 days (log t_o = -1.75) but would have a disruption lifetime of 7 years (log t_2 = + 0.85).

Fig. 2 is a log plot of t_o versus t_2, with the filled circles and the pluses having their same meaning. In this case, however, it is the pluses which should follow the line of perfect correlation, which they do very nicely, with an rms deviation of only 0.26 in the log. It is the filled circles which should fall below the line of perfect correlation, and again virtually all of them do, many by more than two orders of magnitude. These are mostly the larger spots which would have very long potential lifetimes according to the second law but which are vulnerable to the shear of differential rotation and hence die away much sooner than that. Note the generic sunspot, in the lower left corner of this figure, falling almost exactly on the line of perfect correlation.

Although the filled circles fit the line of perfect correlation in Fig. 1 very well, with almost equal numbers above and below, we point out that t_1 was computed with the minimum disruption times, i.e., with the spots assumed to be at $\phi = 45°$. Had we considered the spots touching the equator or touching the pole, the filled circles would have fallen systematically to the right of the line of perfect correlation, by about 0.3 or 0.2 in the log, respectively.

Remembering the incompletely observed spots for which we had applied the factor 1.5 to obtain t_o, we examined their residuals, $t_o - t_1$ (if $t_1 < t_2$) or $t_o - t_2$ (if $t_2 < t_1$), and found them neither systematically positive nor negative, indicating that our corrective procedure was adequate.

6. Discussion

We think it is remarkable that such a wide range of five observable physical parameters can be united with such a simple two-part law to represent observed starspot lifetimes with an rms deviation less than 0.3 in the log. Note that the ranges covered are 1200× for t_o, 900× for k, 750× for P(rot.), 300× for starspot area, and 80× for R_*.

The two-part law of starspot lifetimes would be fully predictive if one could, at the moment of a given spot's inception, predict what maximum radius it would eventually achieve. With the maximum radius known in hindsight, however, the spot's lifetime is uniquely determined, within a factor of two.

Although the first law is based on the scenario of disruption by differential rotation, and observed lifetimes do obey the first law very well, it is not clear that disruption at the superficial photospheric level is the physical mechanism actually at work. Hall and Busby (1990) suggested a couple of alternative interpretations which could explain the good agreement between observation and disruption theory just as well.

Acknowledgements

This work was partially supported by research grant HRD-9104484 to T.S.U. from the N.S.F. and a travel grant to D.S.H. from the Dunsink Observatory.

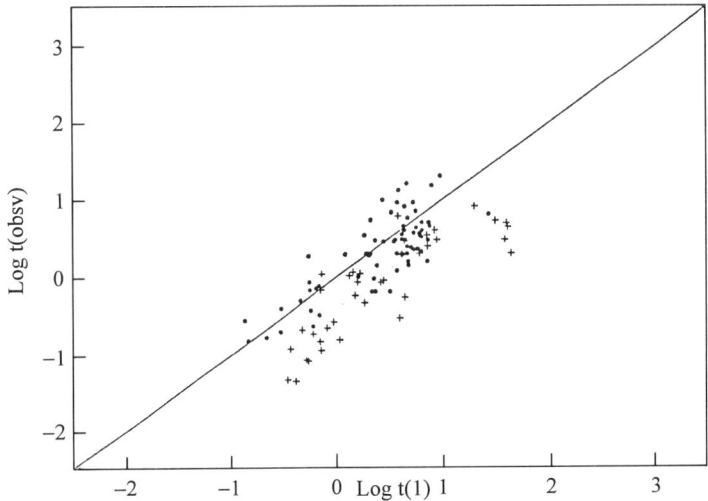

Fig. 1: A log plot of t_o versus t_1, with filled circles indicating spots for which $t_1 < t_2$ and pluses indicating spots for which $t_1 > t_2$. In this case it is the filled circles which should follow the line of perfect correlation, and they do nicely, with an rms deviation of only 0.31 in the log. In this figure the pluses should fall below the line of perfect correlation, and virtually all of them do. These are mostly the smaller spots which do not live as long as their disruption times would permit. Note the small generic sunspot which lives only 6 days (log $t_o = -1.75$) but would have a disruption lifetime of 7 years (log $t_2 = +0.85$).

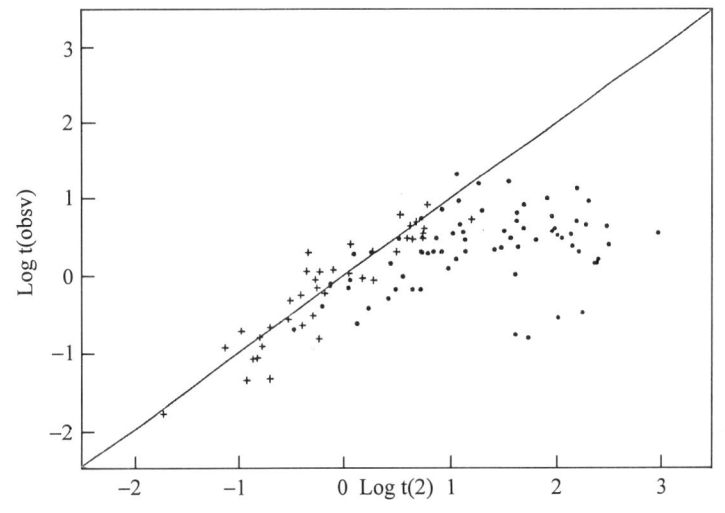

Fig. 2: A log plot of t_o versus t_2, with the filled circles and pluses having their same meaning. In this case, however, it is the pluses which should follow the line of perfect correlation, which they do very nicely, with an rms deviation of only 0.26 in the log. It is the filled circles which should fall below the line of perfect correlation, and again virtually all of them do, many by more than two orders of magnitude. These are mostly the larger spots which would have very long potential lifetimes according to the second law but which are vulnerable to the shear of differential rotation and hence die away much sooner than that. Note the generic sunspot, in the lower left corner, falling almost exactly on the line of perfect correlation.

References:

Allen, C.W., 1973, *Astrophysical Quantities,* London: Athlone.
Gray, D.F., 1988, *Lectures on Spectral Line Analysis of F, G, and K Stars*, Arva, Ontario.
Hall, D.S. and Busby, M.R. 1990, in: *Active Close Binaries*, ed. by C. Ibanoglu, Dordrecht: Kluwer, p. 377.
Hall, D.S., Fekel, F.C., Henry, G.W. and Barksdale, W.S., 1991, *A.J.*, **102**, 1808.

On-going studies of R CrB and UU Her stars with robotic telescopes

J.D. Fernie

David Dunlap Observatory, University of Toronto, Canada.

Abstract

The author has had UBV programs on UU Herculis stars, R CrB, and the chromospherically active δ CrB running on the Automatic Photoelectric Telescope Service. These programs are now in their seventh year, and a brief survey is given of some of the results to emerge from the thousands of observations obtained so far.

1. Introduction

R CrB stars are well-known for their unpredictable habit of suddenly fading in brightness by as much as six magnitudes or more on a time-scale of a few weeks, and then staggering back in fits and starts towards their normal brightness over the following months. The clues to this remarkable behaviour are twofold: first, the R CrB stars are highly evolved supergiants of low surface-gravity, and second, their outer regions have the remarkable property of being almost devoid of hydrogen but exceedingly rich in carbon. The low gravity means that matter can be shed by the star rather easily, while the high carbon content means that most of the matter lost will be carbon. It is thus believed that the fading of the R CrB stars does not presage any major changes in the star itself, but is produced by expelled carbon condensing into soot and veiling the star. The question remains as to what causes the expulsion of the carbon cloud?

It has been known for some time that in addition to their massive but probably extrinsic fadings, most—perhaps all—R CrB stars are pulsationally variable with periods of order 40 or 50 days and amplitudes in yellow light of a few tenths of a magnitude. This pulsation is complex, often with varying periods and amplitudes, and possibly with more than one period present at a time. It has been mooted (Goncharova et al. 1983 and references therein) that the trigger for the carbon expulsion and subsequent deep decline in light lies in the pulsation mechanism.

UU Her stars, although less well-known than the R CrB ones, share a number of their properties. They are F-supergiants superficially similar to normal Population I F-supergiants, but, unlike the latter, are found at significant distances from the galactic plane. This similarity initially led to the assumption that they must be quite young objects and, in turn, this posed the problem of how they could have reached such a location in so short a time. More detailed studies now suggest that at least most of the UU Her stars (UU Her itself may be an exception!) are low-mass, evolved stars, similar to their R CrB brethren, but lacking the low-hydrogen, high-carbon content of the latter, and thus not undergoing deep declines in light. The similarity is strengthened by the fact that the UU Her stars also undergo irregular pulsations of the period and amplitude ranges seen in R CrB stars.

A clear understanding of the type of pulsation taking place in both the R CrB and the UU Her stars would be a useful tool in elucidating the structures of these objects, and if secular period changes were to be found, the direction and speed of evolution could be established. In particular, a detailed study of the pulsational frequencies present in R CrB stars is needed to test the Goncharova hypothesis. Finally, if a single-frequency episode in a UU Her star could be studied in detail both photometrically and spectroscopically, a Baade-Wesselink analysis (albeit a rough one) could yield important clues to the star's basic parameters.

2. Observational work

Studies of this kind require dense observational coverage of the stars, and robotic telescopes in good climates are ideally suited to this. About seven years ago I started just such a project through the Automatic Photoelectric Telescope Service in Arizona, obtaining UBV photometry of R CrB itself and two UU Her stars, V441 Her (89 Her) and V814 Her (HD 161796) at the rate of an observation per clear night throughout each star's annual season. The programme will continue for at least ten years in all, and culminate in an analysis of the accumulated data from the standpoint of the questions posed above. The data are meanwhile being published at about two-year intervals (e.g., Fernie and Lawson 1993, Fernie and Seager 1993, and references therein).

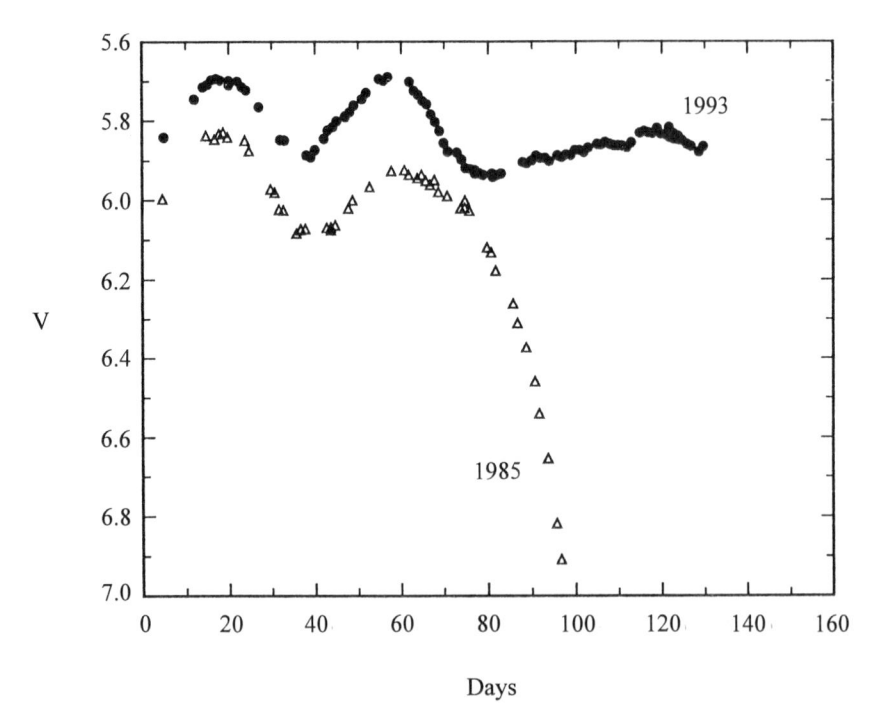

Fig. 1: The light curve of R CrB at the start of two deep declines. The two data sets have been offset slightly from one another in both magnitude and time to better illustrate the differences between them. Note that the pulsation phase at which the deep decline begins is very likely different between the two cases.

As a sample of the data so far obtained, Fig.1 presents observations of R CrB at two epochs when it was at the start of deep declines, illustrating the quite different appearance the light curve can take on such occasions.

References:

Fernie, J.D. and Lawson, W.A., 1993, *MNRAS*, **265**, 899.
Fernie, J.D. and Seager, S., 1993, *PASP*, **105**, 751.
Goncharova, R.I., Kovalchuck, G.U. and Pugach, A.F., 1983, *Sov. Astrophys.*, **19**, 161.

Corner stars for calibration of the Sky Survey fields—results from a pilot programme

P.W. Hill[1], P.O'Neill[1] and B.M. Lasker[2]

[1]*Department of Physics and Astronomy, University of St Andrews, Scotland.*
[2]*Space Telescope Science Institute, Baltimore, USA.*

Abstract

A basic photometric calibration for Schmidt plates is proposed based on stars in the centres and overlapping regions of all Survey fields. A pilot study with the St Andrews Twin Photometric Telescope has demonstrated the feasibility of the project for such a telescope operating under automatic control.

1. Introduction

The reference stars available for photometric calibration of wide-field images, e.g., the major atlases of Schmidt Sky Survey plates, are generally concentrated in small areas typically at the plate centre (cf., e.g., Lasker et al. 1988, Postman et al. 1992, Maddox and Sutherland 1992). In such cases, while the quality of the photometry near the sequences is excellent, at distances of the order of degrees therefrom the effects of errors due to photographic non-uniformity greatly exceed the random errors of measuring and fitting to the reference stars. (Vignetting need not be included in this discussion as it is a measurable, albeit spatially complex, function.)

The photometric issues in question affected the Guide Star Catalogue for the *Hubble Space Telescope* (GSC, Lasker et al. 1990, see also error discussion in Russell et al. 1990). As several GSC maintenance programmes, including the incorporation of new northern data based on the POSS II plates (Lasker 1992), are in progress, the natural possibility of a new photometric calibration effort dedicated to decreasing the field effects motivates the present collaboration. We therefore consider a grid of photometric standard stars to provide zero points at the centre and corners of each Sky Survey field. While the small overlap between the POSS I fields requires a large number of stars to tie together securely the magnitudes on adjacent plates, the situation is much more favourable for the modern surveys taken with $6.4°$ plates at $5°$ centres such as the various Atlases from the UK Schmidt (Morgan et al. 1992) and the Second Palomar Observatory Sky Survey (POSS II, Reid 1988).

2. Corner stars

In a simple Cartesian grid of $6°$ square fields at $5°$ centres there will be a 1 sq$°$ overlap of four fields in each corner (Fig. 1(a)), so that one centre and one corner star would be required for each of the 894 survey fields in each hemisphere. However, the standard application of such a grid on the spherical sky (Tritton 1983; a mirror-image of that grid is used in the north) shows that the ideal case only applies to declination zones within $15°$ of the equator. As at higher declinations the situation is more complex, most overlap regions being of three fields, not four (Fig. 1(b)), we estimate that an additional 384 corner stars would be required in each hemisphere. Detailed planning, based on mapping the GSC onto

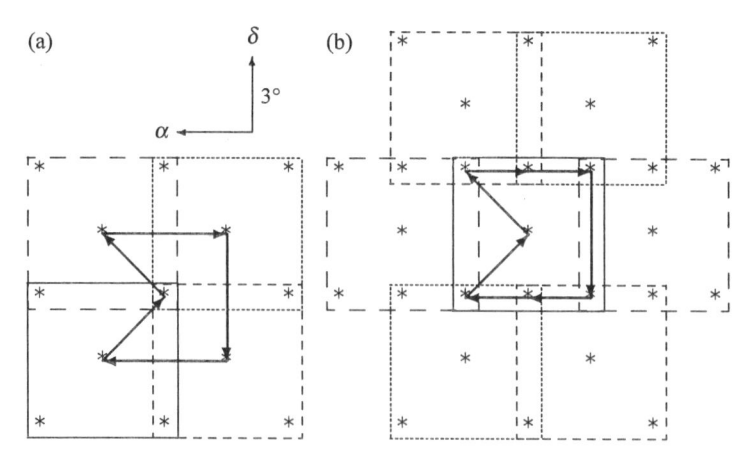

Fig. 1: Schematic representation of Schmidt Survey field overlap regions and observing sequences: (a) four-field overlap with sequence based on corner star, (b) three-field overlaps due to displacements between declination zones and sequence based on centre star.

the plate grid, will be required to identify suitable stars. The basic observational strategy would be to compare each centre star with its surrounding corner stars, possibly followed by comparisons of each corner star with its surrounding centre stars (Fig. 1(a)). This is an ideal task for a robotic telescope.

3. The proposed programme

The Twin Photometric Telescope (TPT) was designed for simultaneous differential stellar photometry. Originally at the Royal Observatory, Edinburgh (Reddish 1966), it was moved to the University Observatory, St Andrews, and used for a programme of observations of eclipsing binaries (Bell and Hilditch 1984) which showed that differential magnitudes could be determined with a precision of much better than ± 0.01 mag even in much less than perfect photometric conditions. The TPT consists of two 40-cm Cassegrain reflectors on a single-fork mounting, the offset telescope being movable by up to $5°$ in both RA and Dec relative to the reference telescope. The telescopes have recently been fitted with new programmable photometers with aperture offsets for sky measures and a CCD TV acquisition system (Edwin and Gears 1992). The drive and control systems are currently being upgraded to fully programmable operation (Edwin and Gears 1993).

A Sky Survey field is observed by setting the reference telescope on the centre star and cycling the offset telescope around the corner stars (Fig. 1(b)). Control of the relative sensitivity of the two photometers is maintained by starting and ending each sequence with both telescopes on the centre star. Filter sequences are observed for each pair and appropriate sky measures included. Parameters will be adjusted to keep systematic errors less than 0.02 mag and random errors to less than 0.05 mag to match the precision of the best photographic magnitudes achievable (but desirably to 0.02 mag). The largest contribution is likely to be from errors in differential extinction at high air mass which should be restricted to less than 2.0. Photometric standards (e.g., Landolt 1983) will be included in the sequences as appropriate.

4. Choice of corner stars

As POSS II includes blue and red plates it is desirable to measure in the B, V and R bands. The stars should be as faint as the TPT can effectively support, say $V =$ [11.0,12.0]. Stars in this range are relatively easy to measure on the POSS II plates and have a reasonable surface density (Bahcall and Soneira 1980, give 5.5 stars/sq° at $V = 11$ at the Galactic pole), while brighter stars are of lesser interest as they are likely (at least in B and V) to be in the Tycho catalogue (Egret et al. 1992).

We also note that a natural complementarity exists between this project and the grid that will come from the Tycho observations (Høg et al. 1992, Egret et al 1992), the major difference being one of magnitude distribution, for at the magnitude range of this project, the surface density of Tycho stars is prohibitively low and the photometric precision insufficient.

Candidates in the overlap regions will initially be selected from the GSC. Ideally, main sequence F and G stars are the most suitable. Catalogues and particularly the INCA (Turon et al. 1992) and SIMBAD databases will be searched to eliminate unsuitable stars such as variables and binaries. Very early and late spectral types and stars with IR excesses would also be eliminated as being photometrically unpredictable. However, it is likely that there will be no information on the majority of stars initially selected and the observing programme will be designed to detect variables as far as is possible.

5. The pilot programme

For the pilot study observations were made during 1992 January and February in the four adjacent POSS I fields in the region $4^h 12^m < \alpha < 5^h 00^m$, $+9° < \delta < +21°$ Because of commissioning problems with the cooling of the acquisition TV CCD (Edwin and Gears 1992) this initial study was restricted to SAO Catalogue stars with visual magnitudes between 6 and 9 and to the V filter only. Known binaries, variables, late K and M stars, and stars in the IRAS Point Source Catalogue were rejected from the selection. Observations in each field were made as described above, sky measures being made in both telescopes after each observation. Integration times were multiples of 10 s, up to 3 min in some cases, on the stars and up to 50 s on sky.

The data were reduced using computer programs written for the reduction of long sequences of variable star observations obtained with the TPT. Although not entirely suitable for the present purpose, they included corrections for differential extinction and zero-point differences between the telescopes, determined from the centre—centre observations. The differential magnitudes for each field are given in Table 1 in the sense centre—corner, the errors being standard deviations for a single 10-s integration. V_{cat} is the visual (in parentheses) or photoelectric V magnitude from the SIMBAD database. At least one, but not always the centre, star in each field had a photoelectric V magnitude enabling magnitudes V_{Obs} to be determined for the others, but no colour corrections have been applied to these data which are only suitable for illustrative purposes. In all but one case the residuals between the catalogue and observed differential magnitudes for those stars in each field with photoelectric magnitudes are gratifyingly small, the mean being +0.023 mag ± 0.022 mag (se). Two stars, SAO93928 and 76548, were designated as variables subsequent to the fourth edition of the GCVS and therefore not eliminated from the preliminary selection. Published V magnitudes for the former show no evidence of significant variation and are supported by

Table 1— Observations with twin photometric telescope

SAO	ΔV	σ	V_{Cat}	V_{Obs}	SAO	ΔV	σ	V_{Cat}	V_{Obs}
\multicolumn{5}{c	}{Field 1501}	\multicolumn{5}{c}{Field 1504}
94017	0.00		8.06		94000	0.00		(8.6)	8.68
93928	−0.59	0.01	7.49	7.47	111759	−2.15	0.01	6.53	
94074	+0.02	0.01	8.09	8.08	111952	−0.49	0.02	(9.2)	8.19
76693	+2.51	0.01	(9.7)	10.57	94097	−0.67	0.01	(7.8)	8.01
76548	−1.55	0.01			93903	−1.20	0.02	7.60	7.48
\multicolumn{5}{c	}{Field 1481}	\multicolumn{5}{c}{Field 1534}
94203	0.00		(9.1)	9.22	94193	0.00		7.14	
94114	−0.89	0.01	8.33		112027	+2.20	0.01	(8.6)	9.34
94302	−1.48	0.01	(7.8)	7.74	112360	+0.50	0.02	7.59	7.64
76923	−0.78	0.01	(8.3)	8.44	94322	+0.61	0.01	7.76	7.75
76767	−2.22	0.03	7.03[a]	7.00	94116	+1.05	0.01	(8.2)	8.19

[a] V from Oja (1987).

our result. Because of the large number of additional observations required no attempt was made to link together these adjacent but non-overlapping POSS I fields.

6. Feasibility study

The pilot programme was undertaken in order to attempt an estimate, however crude, of the time the main programme might take to observe the entire northern sky. There are a number of factors to consider:

1. The number of individual observations required:

Because of the displacement of the field centres between declination zones discussed in Section 2 above, the average number of corner stars in any one field is increased by 21% to almost 5. We assume 5 centre—corner observations and 2 centre—centre observations in a sequence on each field.

2. The time per observation:

Repeated 10-s integrations on the faintest star observed in the pilot programme, $V = 10.57$, have standard deviation ±0.014 mag. For a star 1 magnitude fainter we estimate the error to be 0.03 mag which will be reduced to <0.02 mag if all fields are covered once based on centre stars. The total integration time for three filters may therefore be estimated conservatively as 60 s to include sky measures. The time required to set and centre the offset telescope depends very much on the efficiency of the automation process including field identification. The present drive motors take 25 minutes to traverse around the 5° square, however many corner stars are visited on the way. Allowing 14 minutes per sequence for acquisition and integrations and 5 minutes to set the reference telescope on the centre star makes the total time about 45 minutes per sequence. We consider this to be unacceptably slow and are currently investigating ways of speeding up the offset telescope motions by a factor of 4 or 5, so that we can aim at three sequences per hour for the automated telescope. This would enable the programme based on the centre stars to be completed in 300 hours observing.

3. The number of possible nights in an observing season:

At the latitude of St Andrews (+56° 20') photometric observations are only possible from September through April, which can make for difficulty completing observations on the summer sky. Some observing is possible on about 40 to 50 nights a year on average. If the average length of night is, say, 8 hours, it would theoretically be possible to complete the programme within one observing season. However, two, three, or even more years might be a more realistic estimate and complementary observations from a site having a more favourable summer season may be required.

4. The time available for this project:

The estimate above supposes that the telescope would be dedicated to this project and all possible observing time used. Other programmes would be expected to continue, and the project depends on the availability of manpower. The major advantage of automation in a north European climate is to increase efficiency and we would not expect the observing to be unsupervised.

7. Discussion

The pilot study has shown that the desirable photometric precision for the corner stars project is easily attainable with the TPT and that, provided sufficient resources are made available, it should be possible to complete the project within a time-scale of a few years, comparable to that of POSS II and the second edition of the GSC.

We do, however, consider it desirable to advance scientific programmes and to demonstrate the viability of the project by placing priority on a subset to a reduced photometric precision of ± 0.05 mag. The selected region will consist of 100 fields in the North Galactic Cap which might be completed in a few good observing nights.

Acknowledgements

P.O'N. undertook the pilot investigation as part of his final-year undergraduate project. We thank Brian McLean for a discussion of the complementarity of this work with the Tycho programme, Dick Gears for instruction and assistance in the use of the TPT and photometers, and Steve Bell for the photometric reduction programs and advice on their use. The St Andrews STARLINK node was used for data reduction. This research has made use of the SIMBAD database, operated at CDS, Strasbourg, France. The Space Telescope

Science Institute is operated by The Association of Universities for Research in Astronomy, Inc., under contract to NASA.

References:

Bahcall, J.N. and Soneira, R.M., 1980, *Astrophys. J. Suppl.*, **44**, 73.
Bell, S.A. and Hilditch, R.W., 1984, *Mon. Not. R. astr. Soc.*, **211**, 229.
Edwin, R.P. and Gears, R.T., 1992, *Publ. astr. Soc. Pacif.*, **104**, 1234.
Edwin, R.P. and Gears, R.T., 1993, in: *Stellar Photometry—Current Techniques and Future Developments, IAU Coll. No. 136* (Poster Papers), eds. Butler, C.J. and Elliott, I., Dunsink Observatory, Dublin, p. 36.
Egret, D. et al., 1992, *Astr. Astrophys*, **258**, 217.
Høg, E. et al., 1992, *Astr. Astrophys*, **258**, 177.
Landolt, A.U., 1983, *Astr. J.*, **88**, 439.
Lasker, B.M., 1992, in: *Digitized Optical Sky Surveys*, p. 87, eds. MacGillivray, H.T. and Thomson, E.B., Kluwer, Dordrecht.
Lasker, B.M., Sturch, C.R. et al., 1988, *Astrophys. J. Suppl.*, **68**, 1.
Lasker, B.M., Sturch, C.R., McLean, B.J., Russell, J.L., Jenkner, H. and Shara, M.M., 1990, *Astr.J.*, **99**, 2019.
Maddox, S.J. and Sutherland, W.J., 1992, *ibid*. p. 53.
Morgan, D.H. et al., 1992, *ibid*. p. 11.
Oja, T., 1987, *Astr. Astrophys. Suppl.*, **71**, 561.
Postman, M. et al., 1992, in: *Digitized Optical Sky Surveys*, p. 61, eds. MacGillivray, H.T. and Thomson, E.B., Kluwer, Dordrecht.
Reddish, V.C., 1966, *Sky and Telescope*, **32**, 124.
Reid, N., 1988, in: *Mapping the Sky—Past Heritage and Future Directions, IAU Symp. No. 133*, p. 331, eds. Debarbat, S., Eddy, J.A., Eichhorn, H.K. and Upgren, A.R., Kluwer, Dordrecht.
Russell, J.L., Lasker, B.M., McLean, B.J., Sturch, C.R. and Jenkner, H., 1990, *Astr. J.*, **99**, 2059.
Tritton, S., 1983, *UKSTU Handbook*, Appendix 3, Royal Observatory, Edinburgh.
Turon, C. et al., 1992, *The Hipparcos Input Catalogue*, ESA SP-1136.

The automatic 60-cm telescope of the Belogradchik Observatory—first results

A. Antov and R. Konstantinova-Antova

*Department of Astronomy, Belogradchik Astronomical Observatory,
Bulgarian Academy of Sciences, Tsarigradsko Shose 72, Sofia-1784, Bulgaria.*

Abstract

The Belogradchik Observatory of the Bulgarian Academy of Sciences is situated in the west part of the Balkan mountains at 630 m above sea level. The observatory is equipped with a Carl-Zeiss 60-cm Cassegrain reflector. A one-channel photon-counting photometer with EMI 9789 QB photomultiplier is attached to the telescope.

The automatic system using an IBM-PC/XT computer and synchronized movement of the dome has recently been built. The automation of the telescope includes: RA and Dec. positioning system up to six stellar objects in a field of 6 degrees with accuracy of 5 arcsec each, and automatic moving of the filters; a system for time synchronisation based on the Russian Radio/TV standard with accuracy 0.1 sec; an interface to the computer used for collecting and storing photometric data; a specially created software package "Automatic Photoelectric Reduction" (APR), written in Turbo-Pascal 5.0 for a PC/MS-DOS 5.0 system. This software package contains programmes for data reduction in the standard UBV system and visualisation of the light curves.

Two types of observations are usually run: an estimate of the photoelectric magnitudes of the stellar objects, and patrol observations. The minimum possible integration time is 0.1 s and the maximum is not limited. The automatic 60-cm telescope of the Belogradchik Observatory is thus suitable for observations of stellar objects, asteroids, galaxies and for searching and detecting very rapid brightness variations.

1. Introduction

The Belogradchik Observatory together with the Rozhen Observatory are both observational stations of the Department of Astronomy, Bulgarian Academy of Sciences. It is situated in the west part of the Balkan mountains, 180 km from Sofia, near the well-known rock formations of the small town of Belogradchik. The co-ordinates of the observatory are: E. Long.: $+22°40.1'$; Lat.: $+43°37.6'$. The altitude above the sea level is 630 m. There are approximately 180 clear nights per year.

The observatory is equipped with a Carl-Zeiss 60-cm Cassegrain reflector with equivalent focal length 7.5 m. The one-channel photon-counting photometer with EMI 9789 QB photomultiplier, attached to the telescope and a set of six diaphragms (from 13.8 arcsec 0.5 mm to 137.5 arcsec = 5 mm) were built at the workshop of the Department of Astronomy in Sofia. The UBV and *uvby* photometric systems are available. The respective filters are mounted in a common wheel. The *uvby* system and the associated software are described in Kalcheva et al. (1991).

The automatic system includes an RA and Dec. positioning system, a system for time synchronization, an interface to the computer for collecting and storing data and the software package "Automatic Photoelectric Reduction" (APR) for data processing.

Fig. 1: Belogradchik Astronomical Observatory in December 1993. The 60-cm telescope is housed in the dome on the right. A new 36-cm Celestron has recently been installed in the left-hand dome.

2. RA and Dec. positioning system

The system consists of stepper motors, a special interface with RA and Dec. control and automatic movement of the filters. The software is written for IBM-PC/XT/AT computers in CI 5.0. A more detailed description of the system is given in Staikov et al. (1994). The system allows positioning up to six stellar objects in a field of 6 degrees with accuracy of 5 arcsec. There are two ways to run the system: from a control panel of the telescope or from the control room, using the computer keyboard.

Recently an automatic system, using an IBM-PC/XT computer and synchronized movement of the dome have been installed. The system is shown in Fig. 2.

3. Time synchronisation system

The computer clock is used for measuring the time. At the beginning of 1992 a special system for time correction was built. The system is based on the Russian Radio/TV 14 MHz standard with an accuracy of 0.1 s. Every minute corrections of the clock are made.

4. Interface to the computer used for collecting and storing photometric data

The interface includes hardware, entrance attenuator, amplifier-discriminator, photon counter, registers and decoders (Staikov et al. 1994). Dialog control software, written in CI 5.0 for IBM-PC/XT/AT computers is created.

BELOGRADCHIK ASTRONOMICAL OBSERVATORY

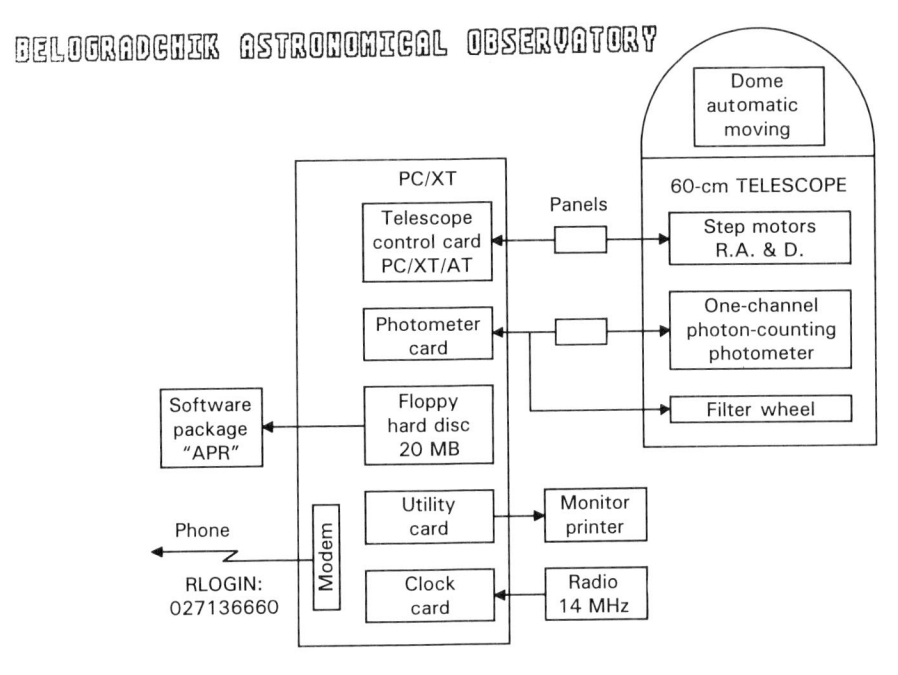

Fig. 2: Automatic 60-cm telescope system

The minimum integration time is 0.1 s, the maximum is not limited. The number of measurements is also unlimited. It depends on the storage capacity of the hard disc only. In our case it is 20 MB. During monitoring all the data up to 640 points are displayed on the monitor in real time. There is the capability of writing every data point to the hard disc during the measurement. The software produces files, containing information about the number of every measurement, UT, the filter used, the value of every measurement, and short indication of what kind of object is measured (variable star, comparison star, background etc.).

Two types of observations are usually run: estimates of the stellar magnitude and monitoring (patrol observations).

5. The software package "Automatic Photoelectric Reduction" (APR)

This specially created package, written in Turbo-Pascal 5.0 for PC/MS-DOS 5.0 systems contain programs for data reduction to the standard UBV photometric system and visualisation of the light curves in the instrumental system (Kirov et al. 1991). The method of differential photometry has been applied. The reduction equations used are (Hardie 1962, and Jerzykevicz 1986):

$$\Delta V = \Delta v - k_v \cdot \Delta X + \varepsilon \cdot \Delta(B - V)$$

$$\Delta(B - V) = \mu \cdot \Delta(b - v) - \mu \cdot k'_{bv} \cdot \Delta X - \mu \cdot k''_{bv} \cdot \Delta(b - v) \cdot \overline{X}$$

$$\Delta(U - B) = \psi \cdot \Delta(u - b) - \psi \cdot k'_{ub} \cdot \Delta X - \psi \cdot k''_{ub} \cdot \Delta(u - b) \cdot \overline{X}.$$

Here Δ is the difference between the variable and comparison star values; $\overline{X}$ is the average mass for two stars; X is the airmass for a star; k_v, k'_{bv} and k'_{ub} are the first-order extinction coefficients, and k''_{bv} and k''_{ub}, are the second-order extinction coefficients, and ε, μ, ψ are the transformation coefficients from the instrumental to the standard UBV system. In 1992 the above-mentioned coefficients have the following values:

$$\varepsilon = -0.093; \; \mu = 1.166; \; \psi = 0.945; \; k''_{bv} = 0\ 04; \; k''_{ub} = 0.08.$$

There are four possibilities for approximation: mean value; nearest time point value; linear approximation; parabolic approximation. Some characteristic statistics (mean value and dispersion) can be calculated and displayed numerically and graphically too.

6. First observations

The automatic 60-cm telescope at the Belogradchik Observatory is suitable for searching for and detecting very rapid brightness variations and for observations of stellar objects, asteroids and galaxies down to $V = 13.5$ mag.

A programme for investigation of fast flare events (few seconds duration) was started at Department of Astronomy, Bulgarian Academy of Sciences, in 1990. In Fig. 3 a flare of AD Leo (Antov et al. 1991), using the 60-cm telescope at the Belogradchik Observatory is shown.

It was observed almost simultaneously, using also the 60-cm telescope at the Rozhen Observatory on 2/3 February 1990. Both telescopes are identical, except for the automation. The distance between them is 270 km. Another fast flare of the flare star V1285 Aql is shown in Konstantinova-Antova et al. (1992). Collaborative programmes for further investigation of the fast flares are desirable in order to clarify their nature. Another intriguing problem is the flare-like events, detected on some evolved stars off the Main Sequence. We have started a programme for the investigation of these events, using the Belogradchik telescope and the 60-cm and 2-m telescopes at Rozhen since 1991. We began observations of the red giant stars V654 Her and IU Ori, not known to be in multiple systems. In the course of monitoring we found the star SAO 65670, used as check star of V654 Her, flaring. The flares are shown in Fig. 4.

The star was not known to be variable until now, and its UBV colours are close to the late K giants. For the resolution of this problem simultaneous observations and collaborative programmes are desirable too. Another programme at the Belogradchik Observatory is the investigation of the flickering of some cataclysmic and symbiotic variables, including KR Aur, TT Ari, MWC 560 etc. Some photoelectric observations of the star MWC 560 are shown in Tomov et al. (1990).

7. Plans for the future

The automatic 60-cm telescope is the first step on the way to establishing a fully robotic telescope at the Belogradchik Observatory.

A guide CCD, RA and Dec. system for automatic moving to the object from a parked position, and a modem with connection to e-mail, are necessary for the future. In the coming years a CCD for photometry and a two-channel photoelectric photometer are desirable. Equipped in this way the Observatory might participate in the European array of

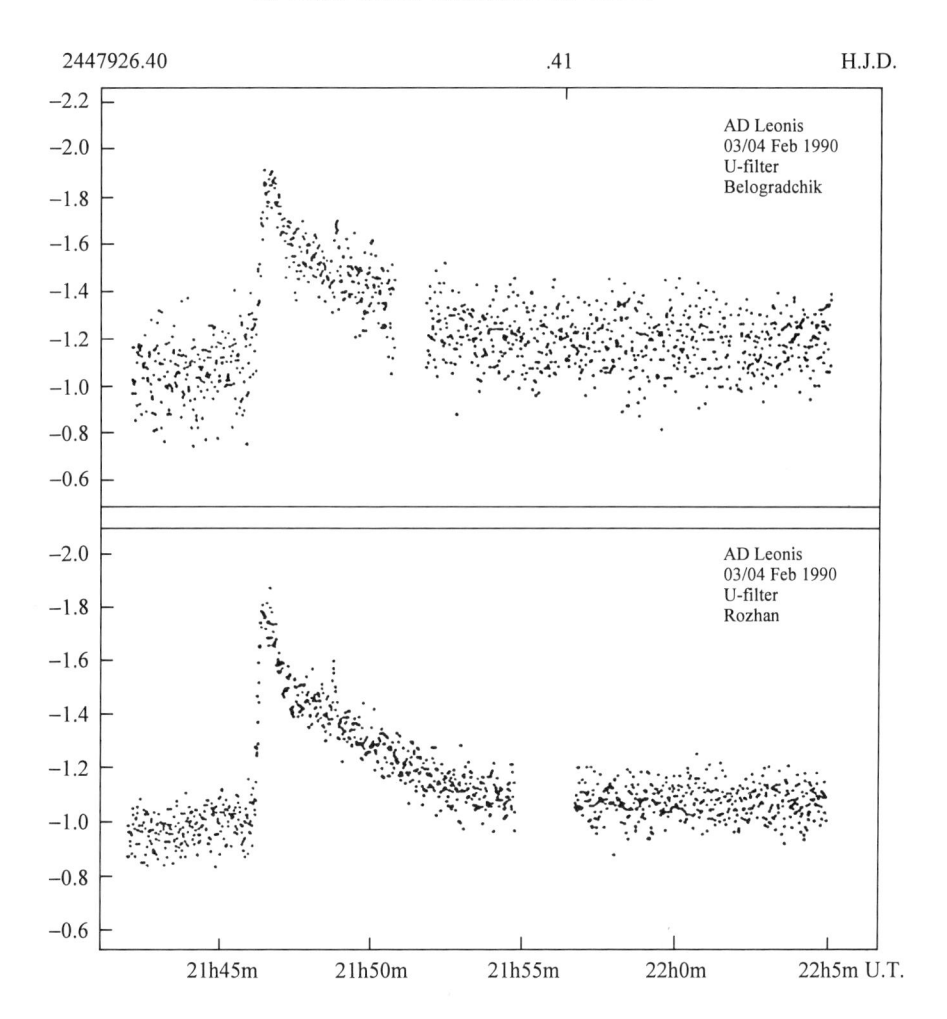

Fig. 3: Flare of AD Leo.

robotic telescopes, to take part in international monitoring programmes for robotic telescopes, simultaneous observations and collaborative programmes. The authors will be very grateful to receive any comments or advice for further development of the automatic telescope.

Acknowledgements

We would like to thank the National Union Club "COSMOS", Bulgaria, for the financial support for building the RA and Dec. positioning system and the interface to the computer for collecting and storing photometer data. The time synchronisation system is built under contract No. MM-58/91 with the Bulgarian National Science Foundation. A.A. is very grateful to Dr. Ian Elliott for support to participate in the IAU colloquium 136 in Dublin and the Kilkenny Workshop on Robotic Telescopes.

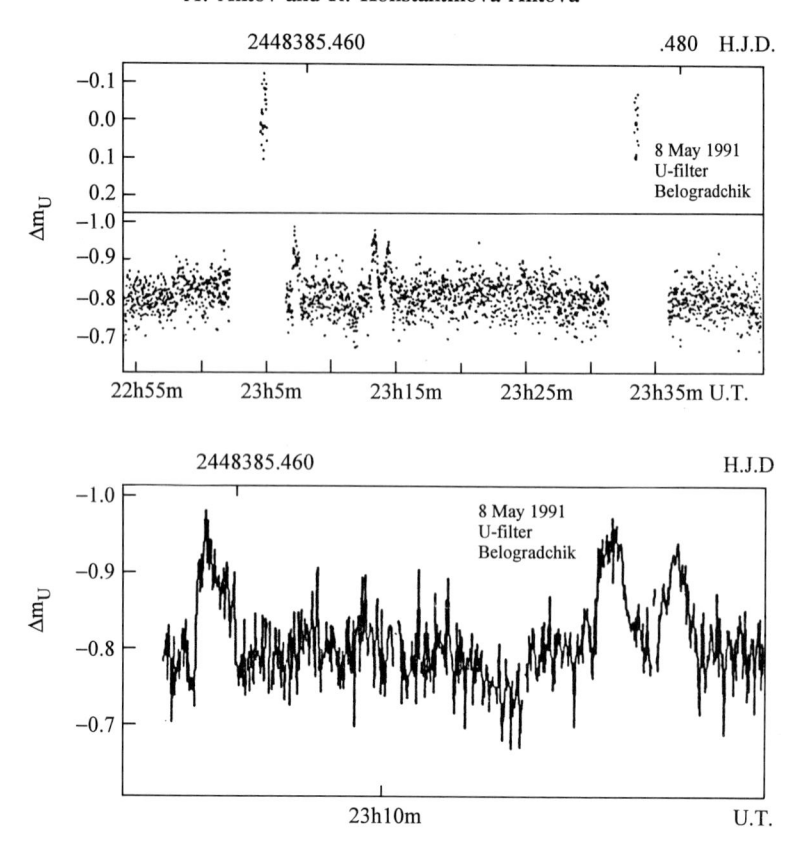

Fig. 4: Flare of SAO 65670

References:

Antov, A.P., Genkov, V.V., Konstantinova-Antova R.K. and Kirov N.K., 1991, *IBVS* 3577.
Hardie, R.H., 1962, in: *Astronomical Techniques*, ed. W.A. Hiltner, Chicago, University of Chicago Press, p. 178.
Jerzykevicz, M., 1986, *Lowell Obs. Bull.*, No. 137, p. 320.
Kalcheva, N., Georgiev L. and Ivanov M, 1991, *Bulgarian Journal of Physics*, **4-4**, p. 17.
Kirov, N.K., Antov, A.P. and Genkov, V.V., 1991, *Compt. Rendus de l'Academie Bulgare des Sciences*, **44**, No. 11, p. 5.
Konstantinova-Antova R.K., Kourdova B.B. and Antov, A.P., 1992, *IBVS* 3695.
Staikov, Yu., Nikov, Ch.N., Sediankov, S.S. and Yaramov, K., 1994 (in preparation).
Tomov, T., Kolev, D., Georgiev, L., Zamanov, R., Antov, A. and Bellas, Y., 1990, *Nature*, **346**, p. 367.

PART III

ROBOTIC TELESCOPE NETWORKS

Some GNAT issues

D. L. Crawford

*Kitt Peak National Observatory,[1]
PO Box 26732, Tucson, AZ 85726, USA.*

Abstract

A global network of automatic telescopes makes sense in a balanced scientific and fiscal program for ground-based astronomy. I explore in this paper some of the issues about new generation small telescopes and a global network, discussing the terminology and the advantages of such systems. An appendix discusses some of the aspects of scaling law estimates of telescope costs as a function of aperture.

1. Introduction

In this contribution, I would like to discuss some of the issues concerning new generation small telescopes and their use in a proposed global network of automatic (or astronomical) telescopes (GNAT). GNAT is a new non-profit, tax-exempt organization, incorporated to provide a mechanism to discuss such issues and to develop a proposal to fund and operate such a facility. Here I will develop some of the background upon which the GNAT concept is based. Please note that this is not a NOAO or KPNO project, in any way.

2. New generation small telescopes

Many of the technological aspects that apply to the new wave of ground-based large optical telescopes also apply new generation small telescopes (NGSTs). Quality of the telescope need not be dependent on the size of the primary mirror, even though the light collecting power varies with the square of the aperture, of course. The potential of such new generation small telescopes is a quantum leap from the performance of older generation small telescopes, just as it is proving to be for the largest telescopes. In astronomy, "small science" is very effective science. Many of the applications to which relatively small telescopes can be put are clearly frontier research programs. Indeed, there are some things that small telescopes can do that large ones cannot.

Let me sketch out some of the aspects of telescopes that we might label "new generation", aspects that are now being incorporated into all of the large telescopes underway or planned as well as into the design and fabrication of many of the newer small telescopes. In fact, it may be useful to define here explicitly what I mean by "small" and "large" in referring to these telescopes. There is, I believe, a natural grouping of telescope sizes, based on several different factors, as set out, for example, in Table 1.

---

[1] Operated by AURA Inc. under cooperative agreement with the National Science Foundation.

Table 1

Size	Aperture (m)	Mirror and Support	"Science"	Cost (USA$)
NGST	0.5 — 1.5	"Easy"	Imaging and photometry	20K — 3M
NGIT	2 — 5	"Do-able"	Higher resolution	3M — 40M
NGLT	6 — 16	State-of-the-art	Very faint, and highest resolution	40M and up

NG, new generation; ST, small telescope; IT, intermediate size telescope; LT, large telescope; OGT, older generation telescope.

Naturally other aspects exist as well, but those in the table are sufficient to give a natural characterization in technology cost, and "science" applications. Clearly some overlap does exist, but the natural divisions do make sense. Another way of thinking about it might be that a NGST can be carried in a pick-up truck and assembled by hand by two people in a small roll-off roof building; a NGIT might fit on a rather large truck (but can be transported easily), and it would need a moderate crane to assemble it, in a special building; while a NGLT needs special transportation in many ways, takes a task force to assemble, and definitely needs a state-of-the-art building with many special features.

Now a few words about what I mean as "new generation" aspects:

Item:	**What is new generation about it?**
Optics	Thin, low weight. Low thermal inertia. Very good images. Relatively fast focal ratios. For NGIT and NGLT: Active optical control.
Mounting	Short, low thermal mass.
Building	Relatively small, simple, low thermal impact on images.
Computer	Fully incorporated into all aspects of the system.
Site	Excellent: many clear observing hours, good seeing.
Operation	Scheduling and operation different from OGTs. Queue scheduling, some automatic operations.
Cost	Relatively low compared to OGTs.
Result	Better performance at lower cost. Value per cost is excellent.

NGSTs, NGITs, and NGLTs share all these advantages. Naturally, some are more important than others, and some are critical for the larger telescopes. For example, it makes no sense at all not to attempt to have the best possible images, limited only by atmospheric seeing (and the site selected so as to minimize that), if one is to have a cost-effective and productive NGLT. Hence, great attention must be paid to any item that may adversely effect the performance of such a telescope.

Another example is cost: as the cost of a telescope scales with approximately the 2.6 power of the aperture, great attention must (or at least, should) be paid to minimizing any factor which will scale at higher powers. For small telescopes, one must also pay attention to items that scale with low power of the aperture, including zero-power. (See Appendix 2 for more details on telescope cost estimating.)

Let me discuss now some of the reasons why small telescopes are very effective (and needed) in astronomy. I will generally use an aperture of 1-meter in such a discussion, both because it is a nice round number and because a 1-meter is a fine example of a size for an effective NGST.

Consider a one meter telescope with a primary focal ratio of about 2 and a secondary focal ratio of about f/10. It will have a field of view (FOV) of about 10 arcmin when used with a CCD of 1000 square pixels, and it will offer 0.5 arcsec/pixel, nearly optimum for photometry under average seeing conditions at a good site. Such matching of pixel size to seeing is actually easier and more natural (and so is the FOV argument) for small telescopes than for large ones.

The 1-meter should be able to do accurate wide band photometry (100 nm band widths approximately), such as UBVRI, for stars of 20th magnitude in exposures of 20 minutes duration. This is better performance than the largest telescope in the world could do only a few decades ago. Such is the power of the new technology. The range of magnitude coverage of a 1-meter telescope is most impressive.

3. Advantages of NGSTs and a GNAT

Let me list now some of the special advantages of NGSTs and a GNAT. These items do not appear in any special order, but flow in some sense from one to another in a stream-of-consciousness way. Such a flow is a natural when considering GNAT and its advantages.

- The value per cost is excellent. While some things can only be done with the largest telescopes, many things can be done with small ones, especially if they are the only ones on which an astronomer can get telescope time.

- The cost of a NGST is relatively low. They are affordable. If we can afford a NGLT, we can certainly afford quite a few NGSTs, even if they are located at relatively remote sites by OGT standards.

- NGSTs are the only hope of getting adequate observing time for many scientific problems and for many astronomers. This is particularly true for the truly "have-nots".

- When located at an excellent observing site (as they usually should be), then they can supply many clear hours of observing (especially when compared to OGTs). Hence:

(i) Time-critical research programs can get done, always.
(ii) Greater-than 12-hour linkage is easy.
(iii) Long-term monitoring is practicable.
(iv) Global coverage can be assured.
(v) Including the northern and southern hemispheres.
(vi) They will be very cost-effective.

- Co-ordinated research is relatively easy, and that will help insure "cross-fertilizations" of ideas. Such efforts will be a great learning process for many of the partners.

- GNAT allows for many users, as there are many telescopes (especially as they are at excellent observing sites). For many users, it may be the only way of getting any telescope time on a good telescope with good instrumentation.

- The impact on those in "Second" and "Third" World countries can and should be very large. They will have access to first-class observing facilities at excellent sites. They will be in easy contact with others interested in similar problems. They will have access to a networked computer system, full of ideas and information and existing data. Research and science education in such locales should benefit greatly.

- NGSTs are very complementary to NGITs and NGLTs, as well as radio astronomy and space facilities. They are a part of a balanced approach to astronomical research.

- While many of the NGSTs can and should be located at excellent observing sites, some can very well be located at less efficient observing sites, closer to campuses or in the home country, and as such be of advantage for educational and for technological developments. Due to the relatively low cost of a NGST, such usage is quite practicable.

- NGSTs make excellent test beds for many aspects of new instrumentation, including detectors and software, as well as new scheduling and observing methods. Thus valuable time on the costly larger telescopes can be optimized.

- There will be a number of aspects of "new science" possible due to the ability to monitor and survey in a manner never possible before, as well as to insure that time-critical events are covered.

- Homogeneous data will be the norm, as most all telescopes, detectors, filters, and observing methods will be nearly identical. Better and more accurate science will result. In addition, there will be the opportunity to get an adequate amount of data (finally), as there never appears to be enough quality time now to do a first-class job. In astronomy, we are beginning to do too much second-class first-class astronomy, and without a new wave of NGSTs the situation will undoubtedly get worse.

[Second-class first-class astronomy = a great science problem with a nicely proposed approach, but one only gets three nights when one needs five, and one of the three is cloudy, and most of another is shot due to equipment problems, and some of the other is lost (or at least not well used) due to the fact that the observers are not familiar with the instrumentation or telescope. So they get marginal (at best) data, and not nearly enough. They publish anyway, of course, because otherwise they will not get any future observing time. So it goes.]

- Coupled or decoupled use of GNAT is quite possible.

- We can develop and put a funded GNAT into operation quite rapidly relative to a new NGLT program.

- Great PR for astronomy, globally and locally.

- Great educational impact:

    (i) Viable and effective extension of university research.
    (ii) Real science with real data.
    (iii) Not the "cold, cloudy night problem"   = A turn-off.
    (iv) It's computers and prime observing sites = A turn-on.
    (v) And excellent data, and enough of it   = A turn-on.
    (vi) One can work when time allows it. A clear night locally is not the issue, nor is the fact that one must work (teach?) in the day and the night is at night. GNAT fits well with academic schedules, and with the fact that most academics are at locales where the observing conditions are not very ideal.

- It can and will be used by users from large universities, small universities, consortia, second and third World institutions worldwide, even high schools and amateur astronomy organizations.

- It is easy to share capital and operational costs with others. $1 + 1 > 2$ can be the norm.

- High-tech and frontier science can be done by all, at low cost.

- It offers facilities and experience and training for going on to more expensive (and therefore scarce) facilities, such as NGLTs or space astronomy.

- A GNAT of NGSTs can and will develop the potentials for networks of NGITs with the type of astronomy they do best (higher spectral resolution than NGSTs). Note also that an array of NGSTs is really a NGIT, perhaps at lower cost even. And so on. It is easy to bubble out the ideas and the potentials of a GNAT.

4. The bottom line: GNAT—why?

- The Value per Cost is excellent, for research and for education.

- Sum > parts.

- All parts are interactive. It is a network.

- It is international. Global. One can observe from anywhere those objects where global or time-critical coverage is needed.

- Only hope for adequate telescope time for many.

- In principle, anyone, anywhere can take full advantage of the system for viable and effective research and education.

- It allows many programs to be done that are difficult now, or impossible: long term monitoring, time-critical observations, backup to space programs, etc.

- All data will be homogeneous. Multiple orders for filters, detectors, even telescopes will become the norm.

- Growth of the network is easy.

Appendix 1. Some definitions.

I often find it useful to look up formal definitions of some of the words we use regularly, for added meaning and to insure that we are using them correctly and not just in some jargon sense. So I list here dictionary definitions of some of the words used above and elsewhere in these issues:

Automatic: Having an inherent power of action. Self-acting or self-regulating. Not voluntary; not depending on the will; mechanical.
Effective: Producing a decided, decisive, or desired effect. Impressive, striking.
Efficacy: Power to produce effects (use of things).
Efficient: Immediately effecting. Highly capable or productive.
Efficiency: Quality of being efficient. Effective operation as measured by a comparison of production with cost in energy, time, or money.
Global: Relating to the globe, especially as an entirety; world-wide.
Network: Any system of lines or channels interlacing like the fabric of a net.
Robotic: Robot: Any automatic device that performs functions ordinarily ascribed to human beings, or operates with seemingly human intelligence. [Robotistic is the adjective.]
Viable: Capable of living. Capable of growing or developing.

Appendix 2. Scaling law considerations.

Consideration of simple scaling law estimates of telescope cost may be a useful tool in some astro-economic discussions. Let us look at a few issues here.

There is no question that the estimated (and actual) cost of a new telescope is an important issue in today's funding environment (or at any time, for that matter). I have discussed the issue with many astronomers and been inspired by earlier work by Rule, Meinel, Disney, and Abt on similar topics.

Consider a general scaling law for telescope costs as a function of aperture, as did Disney and others:

$$\text{Cost} = a_o + a_1 * A^1 + a_{2.5} * A^{2.5} + a_3 * A^3 + a_4 * A^4 \ldots$$

Clearly, this is a general power law expansion that can be used to fit to empirical costs of telescopes with different apertures. Disney discussed how some aspects of a telescope in fact can be estimated as a function of certain powers, at least within certain ranges. Here, however, we will just fit some reasonable telescope costs to one such equation, with two terms only.

In the past, such fits showed that a single term with a power of 2.7 fit existing telescope costs quite well. However, when we consider small telescopes as well as larger ones, it is clear that a two-term equation is required. "Fixed costs" come into play, for items that are rather independent of aperture. One must therefore use at least two exponents. I will here, mainly for simplicity, use a power of "0" and of "2.6" as ones that give a rather good fit to a rational estimate of telescope costs. The lower power term will dominate for the smallest telescopes while the higher power dominates for the larger ones. I give in Table 2 an example of such a fit, with a short description of the columns. I think that the resultant values for costs are rather reasonable. Naturally, existing telescope costs (actual or estimated) have quite a range for any given aperture, but the values in the table are not bad at all for such a simple estimation method. One must be careful, of course, to compare apples with apples. It is not fair to compare an OGT or a special-purpose telescope with a "normal" NGT, of the sense I discussed in the text above. In these figures, I include optics, telescope, building, basic instrumentation, but not site costs.

In addition, I have built on the earlier work by Abt, who discussed the value of small telescopes to astronomy by using scaling laws to fit the number of users, the number of papers, and the number of scientific citations per paper as a function of telescope aperture. Here again, many caveats hold, but the concept is a good one, I think. I have updated some of Abt's data, and I find a lower power than he did to the current data. I have also quizzed a number of astronomers as to their idea of an appropriate power to use, based on their experience. Estimates ranged from negative powers (quite a common answer, in fact) to up to one. None were higher than one. For the other two items, I use the powers estimated by Abt. They appear to make sense to me, if only in an astro-political sense. The number of papers per user is likely to be higher for larger telescopes and the number of citations higher per paper for the larger telescopes as well. It would probably be worthwhile to update Abt's study, with more data and for more recent telescopes, but I feel that the values presented and used here are probably upper limits for any rational values of exponents for these items.

If we set the number of users of a 1-meter telescope per year to be 40 (a bit more than Abt's value, but still low by today's standards), then the number of users of a 4-m telescope would be 80 and 125 for a 10-m telescope (the latter values would be 120 and 300 if we used a power of 1.0). If we assume 320 usable nights, then the average observing run per user would be 8, 4, and 2.6 for the three telescopes, quite reasonable values it appears to me.)

The area of the 4-m is 16× that of the 1-m while the 10-m is 100×. For the same number of photons, sixteen 1-meters would serve more users and produce more papers and more citations than one 4-m, ditto for a hundred 1-meters compared to a 10-m. Granted that the science is different, probably even "more important", but the only way to serve many users is by the viable use of small telescopes.

Table 2—Scaling law example (Cost = 0.13 + 0.22$*$A$^{2.6}$ (unit = M$))

A	A^2	A$^{2.6}$	Cost	A$^{0.5}$	A$^{0.8}$	A$^{1.5}$	A$^{2.7}$
0.50	0.25	0.16	0.17	0.7	0.6	0.4	0.15
0.75	0.56	0.47	0.23	0.9	0.8	0.6	0.46
0.80	0.64	0.56	0.25	0.8	0.9	0.7	0.55
1.00	1.00	1.00	0.35	1.0	1.0	1.0	1.00
1.50	2.25	2.87	0.76	1.2	1.4	1.8	2.99
2.00	4.00	6.06	1.5	1.4	1.7	2.8	6.50
2.50	6.25	10.83	2.5	1.6	2.1	4.0	11.87
3.00	9.00	17.40	4.0	1.7	2.4	5.2	19.42
3.16	10.00	19.92	4.5	1.8	2.5	5.6	22.34
3.50	12.25	25.98	5.8	1.9	2.7	6.6	29.44
4.00	16.00	36.76	8.2	2.0	3.0	8.0	42.22
5.00	25.00	65.66	15	2.2	3.6	11	77.13
8.00	64.00	222.86	49	2.8	5.3	23	274.37
10.00	100.00	398.11	49	2.8	5.3	23	274.37
11.31	128.00	548.28	120	7.0	3.5	38	698.79
15.00	225.00	1142.45	250	3.9	8.7	58	1497.77
16.00	256.00	1351.18	300	4.0	9.2	64	1782.89

The data in the columns have at least the following insights to offer:

A is the size of the light collecting mirror in meters, assumed unvignetted. Note that 2 × 8 m mirrors would give an equivalent mirror of aperture 11.3m. The entry for 3.16 m is included as it gives a collecting area of 10× that of a 1-m telescope.

A^2 is the light collecting power = the area of the mirror.

A$^{2.6}$ is the power we use for the telescope cost scaling law estimate.

Values for A$^{2.7}$ are given for comparison.

The cost of the telescope system is estimated by a two-term scaling law, one-term being independent of the aperture, to include "fixed costs". The second term is the power law, and we assume that the 2.6 power offers the best fit to realistic costs, based either on actual experience or on the best current detailed estimates of telescope cost. The lower power law entries are included so as to facilitate the discussion about telescope "value".

Bibliography:

Crawford, D.L., GNAT, 1992, *A Global Network of Automatic Telescopes*, in: *Automated Telescopes for Photometry and Imaging*, ed. by S.J. Adelman, R.J. Dukes and C.J. Adelman, *Astronomical Society of the Pacific Conference Series*, No. 28, p. 123–127.

Crawford, D.L., GNAT, 1993, *A Global Network of Automatic Telescopes*, in: *Stellar Photometry, IAU Coll. No 136,*, eds. Butler, C.J. and Elliott, I., Cambridge University Press, p. 244.

A complementary network to GNAT: an Arabian and French project for 3T1M automated photometric stations

F.R. Querci,[1] M. Querci,[1] S. Kadiri,[2] and L. de Rancourt[3]

[1]*OMP, Toulouse, France.*
[2]*CNCPRST, Rabat, Morocco.*
[3]*G.I.E. TELAS, Cannes - la Bocca, France.*

Abstract

A project to establish a network of automated photometric stations on very high mountaintops around the north-tropical latitude is in progress. Some Arabian countries are interested in collaborating in this project proposed by French and Moroccan astronomers. The data will be transmitted simultaneously by satellite to all scientific centres of the network. The main scientific aims are the monitoring of variable stars with many characteristic time variations, and the search for planets around stars.

1. Introduction

The prototype of a C-11 twin-telescope has already been described in some papers which also give the first scientific results (Querci and Querci 1986, Fontaine et al. 1987, Querci et al. 1989, Querci et al. 1992). The validity of the equipment has been tested using the algolide, HU Tau and the a δ Scuti star, 63 Her. Short-term oscillations (over about 1 h 30 m) were observed on red giants such as TU CVn. Rapid variations of the atmospheric extinction (over 1 h 18 m) are seen for the Pleiades (C and D stars). Correlation tests on cloud crossing, over the two channels simultaneously, allow amelioration sometimes of the differential measurements (Fontaine 1992).

The 3T1M project was presented at the Joint Commission 9 and 25 meeting during the XXIst IAU General Assembly (Querci and Querci 1992a), when it was a Franco-Moroccan collaboration only. Now, many Arabian countries having high mountains near deserts are interested (Querci and Querci 1992b). This project has also been presented at IAU Colloquium 136 (Querci et al. 1993).

2. 3T1M project principles

Three 1-meter telescopes observe variable, comparison and check stars simultaneously. There is a multi channel photometer on each telescope, hence

- the UBVRI or uvby-Hβ fluxes are recorded simultaneously.
- the star and the sky fluxes are recorded simultaneously (this last point is still under study).

As for reliability, the technology is simple and robust. No technical leap is required. In case of telescope failure, a procedure exists which allows us to continue the observation (which will be less rapid of course). In any case, the principle of simultaneity will be preserved optimally to "freeze" the telluric atmosphere.

3. General specifications

- The station is an automated observatory, i.e. the telescopes as well as the auxiliary equipment work alone. No permanent human presence is required. However, the station will have the capacity to accommodate temporary teams.
- The station is designed for severe environmental conditions.
- The reliability is high.
- The station is self-sufficient for energy production.
- A remote control by satellite high-rate data link (in and out modes) will be provided.

4. Advantages of the 3T1M project

The main advantages are:

- better accuracy than the mono channel technique when the sky conditions are slowly variable:
$$\Delta m_v \approx 0.001 \text{ to } 0.0005$$

- no loss of observing time by telescope manoeuvres and by observations of comparison stars and of the sky background:

    90% of the night time is devoted to the variable

- the angular distance of the three observed objects can be large since the telescopes work independently, i.e. some degrees of arc.

5. 3T1M project initial proposal

In 1991 French and Moroccan astronomers presented a proposal to set a 3T1M automated photometric station at Oukaimeden (altitude ≈ 2870 m; lat. $\approx 31°$N; L $\approx 7°$W) for a scientific collaboration in many fields:

- Stellar variability (asteroseismology) on T Tau, Be, W CMa, δ Scu, Ap and Am stars, red giants and dwarfs, PN etc.

- Planetology: search for planets around stars, stellar occultation in the solar system, etc.

The first drawings were made on mirror characteristics, electronics, satellite transmission optics by: Observatoire de Haute-Provence (OHP), Observatoire Midi-Pyrenees (OMP), Institut des Sciences de l'Univers (INSU), Observatoire de Lyon (OL), in France, plus Laboratoire d'Astronomie et de Geophysique (LAG) in Morocco.

6. Present status

The project is being extended to several sites for an Arabian and French complementary network to GNAT to follow short-time-scale variations. The stations have to be set up on high mountaintops (>3000 m) with latitudes from about $20°$ to $30°$ N and longitudes $7°$ W $<$ L $<$ $120°$ E located under different airstreams and anti-cyclone conditions.

The site selection was begun by the analysis of the meteorological data for some small towns around the site. It progresses via satellite observations through METEOSAT (noting clouds, humidity and sand-bearing winds) and will be definitive after local tests with scintillometers. Some prospective sites are: High-Atlas in Morocco, Hoggar in Algeria, Anatolia in Turkey, north of Lebanon, Sinai in Egypt, north of Yemen, south of Saudi Arabia, east of Oman Sultanate, south of Iran, south of Pakistan, Indonesia. The data analysis of the satellite observations, and those of the local tests, could indicate a minimum number of stations which should be able to pursue observations each night without interruption. It is possible that some additional site prospecting could be made, e.g., in Tunisia, India, the Pamir and Taklamakan deserts.

Relations are being developed with astronomers and/or governments of these countries. The technical participation of laboratories of many countries is under discussion, as well as the financial contribution.

7. Progress and future technical developments

After several months to some years of functioning in the photometry mode only, the network should be progressively equipped for observing spectrophotometrically.

References:

Fontaine, B., 1992, in: *Proc. Workshop on "La Photométrie en France"*, ed. J.P.J. Lafon, Obs. Paris-Meudon.

Fontaine, B., Gregory, C. and Querci, F.R., 1987, in: *Proc. of the Colloquium on "Histoire et Avenir de l'OHP"*, eds. A.A. Chalabaev and M.J. Vin, Observatoire de Haute-Provence/CNRS, p. 191.

Querci, F.R. and Querci, M., 1986, in: *Proc. of the Seventh Annual Fairborn IAPPP Symposium on "Automatic Photoelectric Telescopes"*, eds. D.S. Hall, R.M. Genet and B.L. Thurston, Fairborn Press, p. 156.

Querci, F.R. Querci, M., 1992a, in: *Proc. of Joint Commission 9 and 25 Meeting on Automated Telescopes for Photometry and Imaging*, eds. S.J. Adelman and R.J. Dukes, Jr., XXIst IAU General Assembly, Buenos Aires (Argentina), 23 July—1 August, 1991, Highlights of Astronomy, Vol. 9.

Querci, F.R. and Querci, M., 1992b, in: *Proc. Workshop on "La Photométrie en France"*, ed. J.P.J. Lafon, Obs. Paris-Meudon.

Querci, F.R., Querci, M., Gregory, C. and Fontaine, B., 1989, in: *Proc. of the Tenth Annual Fairborn/Smithsonian IAPPP Symposium on "Remote Access Automatic Telescopes"*, eds. D.S. Hayes and R.M. Genet, Fairborn Press, p. 53.

Querci, F.R., Querci, M. and Fontaine, B., 1992, in: *Proc. of the First European Meeting of the AAVSO on "Variable Star Research: An International Perspective"*, eds. J.C. Percy, J.A. Mattei, and C. Sterken, Cambridge University Press, p. 221.

Querci, F.R., Querci, M., Kadiri, S.and de Rancourt L., 1993, in: Poster Papers from *Stellar Photometry, Proc. IAU Coll. No. 136*, eds. I. Elliott, and C.J. Butler, Dublin Institute for Advanced Studies, p. 122.

Editor's Note:

The updated project, now named the Network of Oriental Robotic Telescopes (the ORT Network), can be found in Proceedings of the Symposium "Robotic Telescopes" held in Flagstaff (Arizona), June 28—30, 1994 (to be published in Pub. Astron. Soc. Pacific Conf. Series, eds. Gregory W. Henry and Mark Drummond), or in Proceedings of the IAU Working Group Meeting on "Problems of Astronomy in Africa", held in The Hague (The Netherlands), August 20, 1994, ed. Allan H. Batten (to be published in *Highlights of Astronomy*, Vol. 10, Kluwer).

Development of an academic network for astronomical use in Bulgaria

P. Delchev[1] and M. Tsvetkov[2]

[1]*Computer Centre of Physics, Bulgarian Academy of Sciences,
Tsarigradsko Shose 72, Sofia-1784, Bulgaria.*
[2]*Department of Astronomy and National Astronomical Observatory,
Bulgarian Academy of Sciences, Tsarigradsko Shose 72, Sofia-1784, Bulgaria.*

Abstract

The first stage of the Bulgarian Academic Research Network (BGARNET) for astronomical use is described. The communications are based on BULPAC—the Bulgarian Government network. The main network control and data collection centers used are in the Bulgarian Academy of Sciences—in the Centre for Informatics and Computer Technologies and in the Computer Centre of Physics. The principal node of BGARNET provides gateways to EARN, BITNET and EUNET and supports e-mail protocol.

1. Introduction

The professional astronomical institutes in Bulgaria work mainly within the framework of the Bulgarian Academy of Sciences (Department of Astronomy) and the University of Sofia (Chair of Astronomy). The observatories—Rozhen National Astronomical Observatory and Belogradchik Observatory—belong to the Department of Astronomy and are situated respectively 250 km SE and 190 km NW of Sofia. Besides these academic astronomical institutes there are a few more amateur stations equipped with small telescopes (Sliven, 60-cm Zeiss reflector; Varna, 50-cm reflector; Gabrovo, 30-cm Celestron reflector; Stara Zagora, 20-cm Coudé reflector, etc.). With the development of networks in Bulgaria, a pilot project for connection of the astronomical institutes with the academic and BULPAC network, was started in 1991.

The Bulgarian Government network BULPAC has been developed since 1990 and its structure includes four main centres in Sofia, Plovdiv, Veliko Tarnovo and Varna (Sabev and Michailov, 1992). The network is based on package transfer according to CCITT X.25 protocol. Through the main server in Sofia connections with DATAEXP (German network) and RADAUS (Austrian network) are realised. BULPAC allows 19.2 kbps exchange of the information and its DTE address is 2841. The capabilities of BULPAC are used from the Bulgarian Academic Research Network.

2. The Bulgarian Academic Research Network

The Bulgarian Academic Research Network (BGARNET) has started to work as a pilot project since the beginning of 1992. It is established as a heterogeneous net on the basis of DEC and IBM computers. The DECnet and SNA protocols are used. Another LAN and PC stations are connected besides the main-frame. The network area is developed on the basis of BITNET/EARN requirements and needs. The international point of support via BULPAC/DATEX-Y is located in the Computer Centre, University of Linz.

BGARNET gives possibilities for e-mail service, ISO File Transfer, Access Management (FTAM) and Virtual Terminal Protocol (VTP) and all other possibilities of the X.25 protocol. Our DTE address is 28412205.

The academic network uses three IBM computers connected through rent lines (9600 bps) and SNA protocol:

- Computer Centre of the Centre for Informatics and Computer Technologies (CICT): IBM 4381, VM5, 16 MB/3.0 GB.
- Computer Centre of Physics: IBM 4341, VM4, 16 MB/2.9 GB.
- Computer Centre of the Institute of Mathematics: IBM 4341, OS/VS2, MVS 3.8 16 MB/8.6 GB.

Another eight DEC computers are connected via rent lines (9600 bps) and DECnet protocol:

- Computer Centre of the Centre for Informatics and Computer Technologies (CICT), VAX 11/750 VMS, 5 MB/0.6 GB.
- Computer Centre of Physics, two VAX 11/730, VMS, 4.5 MB/0.2 GB.
- Computer Centre of the Institute of Mathematics: two DEC station 3100's, UNIX 5.0, 16 MB/0.6 GB.
- Institute of Nuclear Research and Nuclear Energy: μVAX 2000, VMS 4.7, 5.0 MB/0.4 GB.
- Institute of Chemistry, two VAX 11/730, VMS, 4.5 MB/0.2 GB.

On the basis of the BGARNET the co-operative computer network, UNICOM (University Scientific Research Computer Network), was established serving the universities and other scientific institutes in Bulgaria.

Through UNICOM, Sofia University, the Technical University of Sofia, the Central Institute of Scientific and Technological Information (CISTI), the Bulgarian National Library etc. are integrated. Fig. 1 shows the main BULPAC centers in Bulgaria and the links with the astronomical institutes/observatories via UNICOM.

3. Astronomical use of the BGARNET

The astronomical institutions and observatories in Bulgaria are modestly equipped, mainly with the IBM-compatible AT/XT personal computers as follows:

- Department of Astronomy, Sofia, Bulgarian Academy of Sciences: one I486, two I286, two I8086.

  (i) Rozhen Observatory: PDP 11/34A, I486, two I386, two I286 and two I8086.
  (ii) Belogradchik: two I8086.

- Chair of Astronomy, University of Sofia: I486, two I386, and some I286.
- Stara Zagora, Sliven, Varna, etc. public observatories: at least one I8086 each.

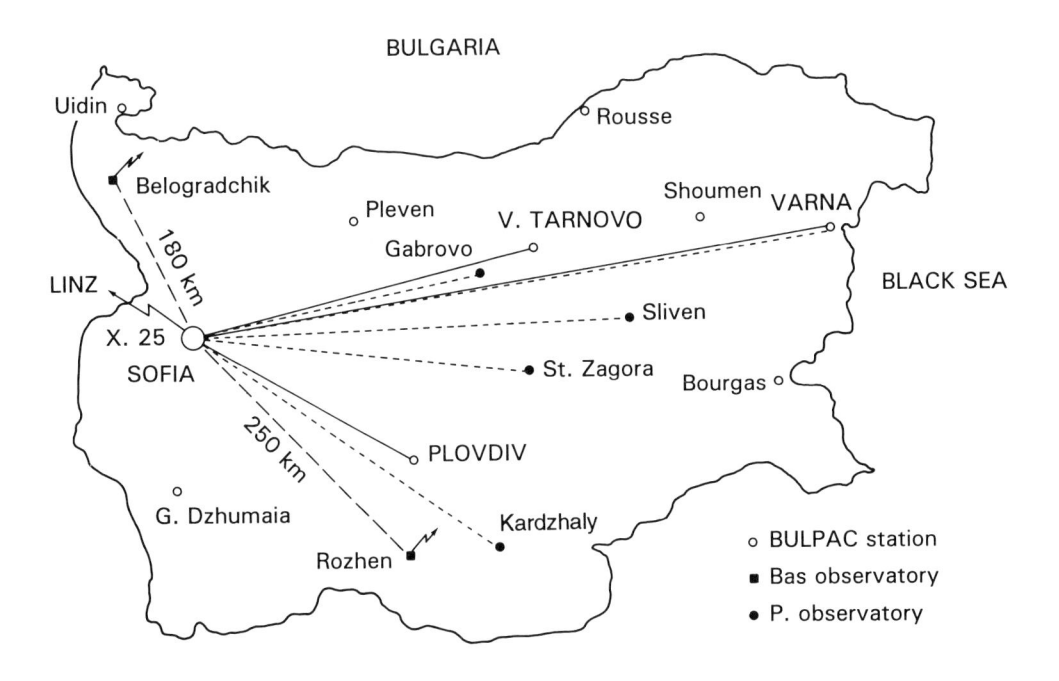

Fig. 1: Main BULPAC centres in Bulgaria and the links with the Astronomical institutes observatories via UNICOM.

The main operating system of the IBM PC is MS DOS and MS WINDOWS. The astronomers from the Department of Astronomy of the Bulgarian Academy of Sciences use the VAX machine of the Computer Centre of Physics, and IBM and VAX of CICT for the network. Their e-mail address is: ASTRO@BGEARN.BITNET. The Chair of Astronomy, University of Sofia uses μVAX 2000 of the Faculty of Physics and their address is: UNIASTRO@BGEARN.BITNET.

Connections by network with the astronomical observatories (Rozhen, Belogradchik, etc.) are in the process of installation using free telephone lines and modems with TELNET protocol (Rlogin: +359 2 713 6660, [2400/1200/300 b/s], MN5, 7.00—19.00 h). The observatories will be connected with main server stations of the academic VAX-computers in the Centre for Informatics and Computer Technology and the Computer Centrr of Physics. Fig. 2 shows a part of the Bulgarian Academic Network for astronomical use.

Through BGARNET we have possibilities for direct file transfer and network links to the European Data Centres (SIMBAD, ESO/STECF - STARCAT, etc.). As a guide in our work on computer communications and data networks in Bulgaria we used the detailed papers of Murtagh (1986) and Boyanov and Sabev (1990).

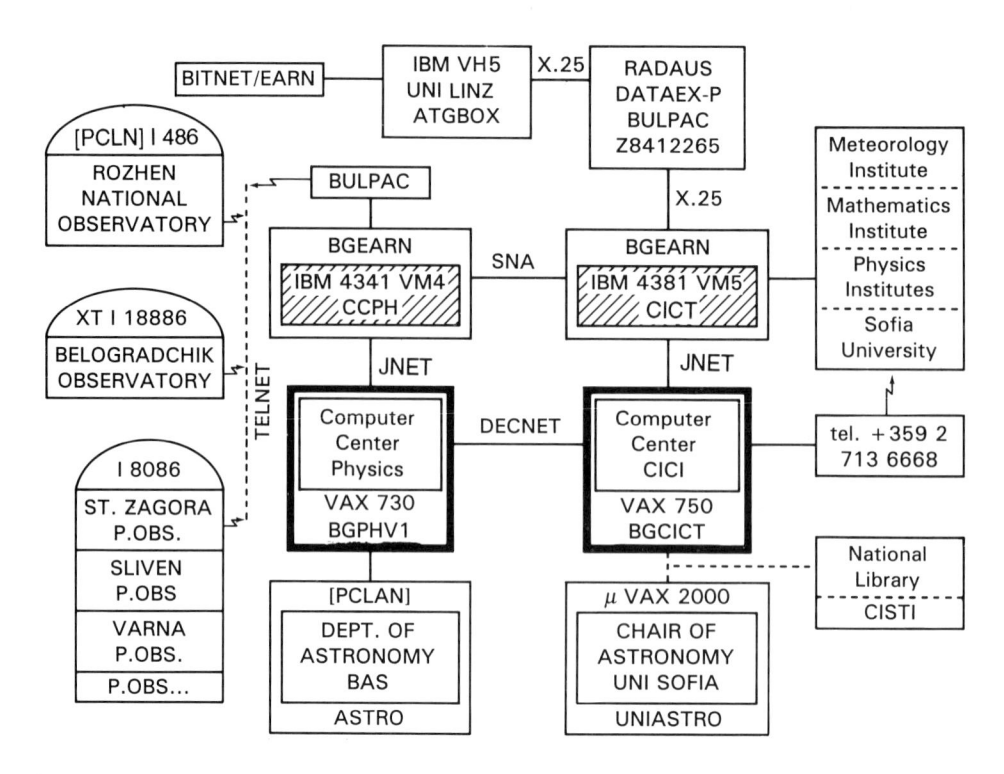

Fig. 2: Bulgarian Academic Research Network for astronomical use (e-mail: astro@bgearn.bitnet)

One of the main current projects based on network use is connected with data collecting and exchange of information for the creation of the database of Wide-field Plate Archives. The Belogradchik Observatory and the server station in the Computer Centre of Physics are in a process of preparation to be included in a network of laboratories across Europe which separately develop precision astrophotometry with robotic telescopes.

4. Future plans

The installation of the LAN in more institutes and observatories is the next stage in the future plans for astronomical use of BGARNET. They will be connected with the global academic network as a sub-network. We plan to use also the UNIX operating system to be connected with INTERNET. In this connection, a special node to INTERNET in the Centre for Informatics and Computer Technologies and the Computer Centre of Physics will be prepared. We plan to be included also in BGNET (a new Bulgarian commercial network), which is connected with EURONET and the main machine "MCSUN" in Amsterdam. From another direction the astronomers will be connected with the National Library and the Central Institute of Scientific and Technical Information, which have access to GEONET in London.

Acknowledgements

We thank our colleagues from the Centre for Informatics and Computer Technologies of the Bulgarian Academy of Sciences and Eng. Peter Ivanov for the help and support of this work. M.T. acknowledges grants and support from Dr. Ian Elliott for participating in IAU Colloquium 136 and the Kilkenny Workshop on Robotic Telescopes.

References:

Boyanov, K. and Sabev, V., 1990, *Automatic Computing Techniques & Automatic Systems*. Sofia, No. 10, p. 38.
Murtagh, F., 1986, *Bulletin Information CDS*, No. 31, p. 89.
Sabev, V. and Michailov, V., 1992, *Computerworld-Bulgaria*, No 26, p. 1.

Facilities for the co-ordination of multi-site and multi-wavelength observing campaigns

C.J. Butler and S.J. Magorrian

Armagh Observatory, College Hill, Armagh, BT61 9DG, N. Ireland.

Abstract

A brief description is given of: (1) a computer program which has been written to assist in the co-ordination of simultaneous observations from ground-based and satellite telescopes, and (2) a newsletter which has been established to notify interested observers of future collaborative programmes.

1. Introduction

Many astronomical studies benefit from the co-ordination of telescopes at different sites, some through the increased time coverage possible and others through the addition of instruments that are sensitive to different wavelength ranges. In recent years it has become increasingly common to make simultaneous observations from satellites and ground-based telescopes, and in most cases the value of the combined multi-wavelength data on a particular phenomenon far exceeds the value of the component parts. A few decades ago, simultaneous radio and optical observations were occasionally made, nowadays it is not uncommon for observations to be carried out simultaneously in X-rays, the ultraviolet, optical and radio regions. The difficulties in co-ordinating such multi-site and multi-wavelength facilities can be quite daunting, particularly to the uninitiated. Here we present details of a computer program designed to facilitate the planning of simultaneous observations using ground-based telescopes and satellites. In addition we describe a simple procedure whereby an observer who already has access to facilities for one wavelength range may contact possible collaborators to provide optical and radio back-up.

2. SCOPES

The computer program SCOPES (Simultaneous, Co-ordinated Observing Program for Earth-based and Satellite-based telescopes), which has been under development at Armagh Observatory in recent years, is designed to help astronomers determine the best time to observe a given list of targets with a given suite of ground-based telescopes and satellites. Up to five ground-based telescopes and five satellites can be employed. The package requires information on the geographical position of the various ground-based telescopes and their observing constraints (e.g., limiting altitude). For satellites the constraints are normally dependent on the position of the Sun, Moon and Earth as seen from the satellite, and may be either absolute constraints (e.g., occultation by Sun or Moon), or warning constraints, where observations may be hampered, but not necessarily forbidden. The package currently incorporates data for many ground-based observatories and for two satellites: IUE and ROSAT. In the case of IUE, the shifts required to observe the targets are given, (i.e. whether ESA, US1 or US2). Other satellites can be easily included, provided their constraints are lunar and solar based. Earth constraints are more complex, particularly for low-Earth orbits. Ground-based radio telescopes are treated differently from optical telescopes in that the restrictions of daylight do not apply.

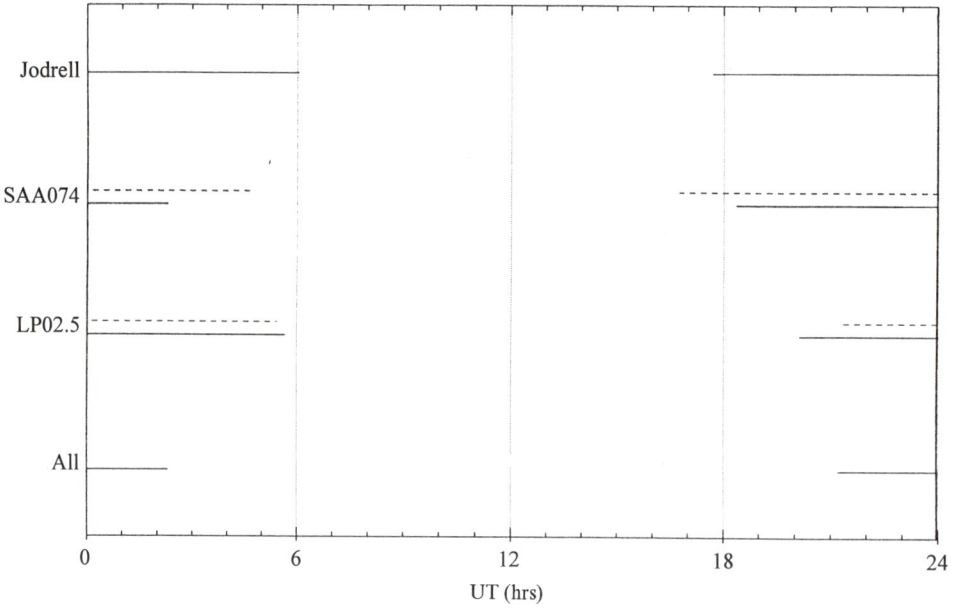

Fig. 1: A time-line showing UT times at which observation of the star Fliese 735 is possible on 10 July 1992, by: a radio telescope at Jodrell Bank, England, the 1.9-m telescope at Sutherland, S. Africa, the 2.5-m telescope on La Palma, Canary Isles, and (bottom) by all three telescopes simultaneously. Solid lines indicate when the star is above altitude 20° and dashed lines, when the sun is below altitude −12°.

Once a satisfactory period for observation has been established, the time-line for various ground-based telescopes can be plotted (see, for example, Fig. 1). In addition, an all-sky map, with an Aitoff projection, can be produced, indicating the relative positions of targets and the various satellite constraints (see Fig. 2). The program uses several of the positional astronomy routines included in the SLALIB package on STARLINK, kindly made available by Dr. Pat Wallace of the Rutherford Appleton Laboratory. For graphical output it requires PGPLOT. Further information on the program, which is written in FORTRAN, can be obtained from the authors.

3. The Co-ordinated Observation Newsletter

Following a meeting of the IAU Working Group on Co-ordinated Observations, held at Baltimore in 1988, a newsletter was established to facilitate observers interested in multi-wavelength and multi-site campaigns. It has been used largely by late-type star enthusiasts, but it is also intended to operate for other sections of the astronomical community, including those interested in extragalactic astronomy.

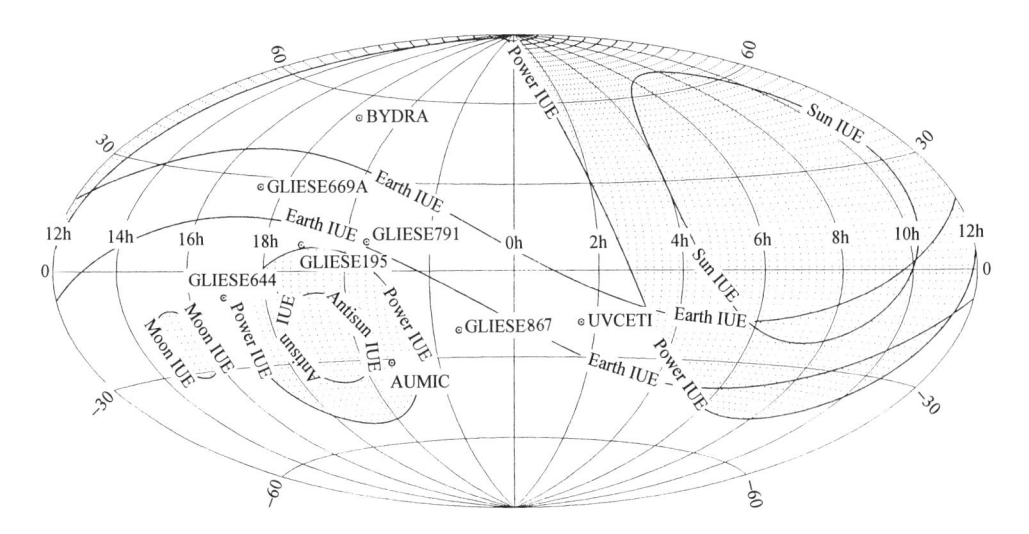

Fig. 2: An all-sky Aitoff projection showing the position of potential targets together with the areas which are constrained by the Sun, Moon and Earth for IUE. The broad band marked "Earth IUE" indicates the path of the Earth as seen by IUE during 24 hours. Shaded areas are forbidden.

Currently, in excess of one hundred astronomers receive the newsletter, which is distributed by e-mail where possible and otherwise by air-mail. It is provided free of charge by Armagh Observatory as a service to the international astronomical community. Astronomers with a scientific programme which requires additional coverage at optical or radio wavelengths, or from sites at different longitudes, are invited to send a contribution to the circular, which should include: (1) a short description of the project, (2) an indication of the additional facilities required, and (3) the name, address and e-mail address of the programme co-ordinator. Contributions, which should not normally exceed one A4 page, should be sent to C. J. Butler at the above address or by e-mail to the INTERNET address CJB@STAR.ARM.AC.UK. Potential collaborators, who wish to receive the newsletter, are also invited to submit their names and addresses to the above.

Acknowledgements

Research at Armagh Observatory is grant-aided by the Department of Education for Northern Ireland.

PART IV

TECHNOLOGY AND IDEAS

Monitoring of active galactic nuclei

I.G. van Breda

School of Cosmic Physics, Dunsink Observatory, Dublin 15, Ireland.

Abstract

Active galactic nuclei vary at all wavelengths, leaving many clues as to their physical nature. Systematic well-sampled monitoring, as can be achieved with robotic telescopes, is therefore important in gaining an understanding of these enigmatic objects. Of particular interest is an attempt to differentiate between the competing black hole plus accretion disc and starburst models. The latter predicts that variable activity will occur due to supernova explosions and remnant activity, which should be observable through photometric monitoring.

1. Introduction

Active Galactic Nuclei (AGN) are well known to be irregular variables at wavelengths from X-rays to radio, with a wide range of time-scales, from tens of seconds in the most active blazars in the X-ray waveband (Turner et al. 1991) to many years. This makes it very difficult to observe AGN following the schedules for time allocation available on most telescopes. AGN are thus ideal candidates for observing with robotic telescopes where observation time can be shared out with other variable objects according to the nature of their variability, rather than in the form of observing runs.

The scale of observational programmes for the measurement of AGN variability is very demanding both of telescopes and of telescope time:

(i) sampling needs to be frequent, around five to ten days in the optical, much more frequent for full X-ray coverage;

(ii) there is considerable scientific benefit to be gained from making simultaneous observations at different wavelengths;

(iii) the objects are faint and need telescopes of reasonable size;

(iv) spectroscopic and polarimetric observations are also very useful and require longer exposure times and/or larger telescopes;

(v) there is a large number of individual objects to observe.

Although the logistic problems are clearly considerable, there have recently been several successful intensive multiple wavelength monitoring campaigns, such as those for NGC 4051 (Done et al. 1990), NGC 4151 and NGC 5548 (Clavel 1991).

2. Models for active galactic nuclei

The problem of understanding the central engine that powers active galactic nuclei is clearly one of the most important aspects of modern astrophysics and has puzzled astronomers for many years. Accepted wisdom has been that AGN are powered by a single central engine consisting of a black hole or, at least, a very massive object surrounded by an accretion disc (Rees 1984).

While the cosmological interpretation of QSO redshifts has occasionally been challenged, particularly by Arp, imaging of nearby QSOs, along with a large body of less direct spectroscopic and other evidence, has indicated that they are phenomena that occur in the centres of galaxies and are just one of a variety of forms of activity that can occur in galaxy environments. Indeed, a veritable menagerie of active extragalactic objects has now been identified, including QSOs, Seyfert I and Seyfert II galaxies, HII galaxies and extragalactic HII regions, starburst galaxies, ultraluminous IRAS galaxies, LINERS, BL Lac and optically violent variables (collectively blazars), with many overlapping properties.

More recently the starburst model, in which the central engine consists of a dense cluster of massive stars, with variable activity due to supernovae and supernova remnants, has come into prominence, particularly through the work of Terlevich and Melnick (1985). This model was originally suggested by Shklovskii (1960), as a means of powering radio galaxies.

2.1 ACCRETION DISC MODEL

In the standard model for radio-quiet AGN, there is an accretion disc surrounding a black hole, with almost all emission coming from thermal processes (Bregman 1991): from dust in the infrared, free-free emission from electrons in the optical, blackbody emission from the accretion disc in the UV and, at X-ray wavelengths, Compton scattering by hot electrons in an atmosphere surrounding the accretion disc.

Blazars and optically violent variables, in general, are considered to have special geometry and to produce non-thermal emission from a jet structure aligned with the observer.

A disadvantage of the black hole models is that they have many free parameters and can be made to fit almost any data. For example, Molendi, Maraschi and Stella (1992) find that the absence of a delay between the UV and optical variability in NGC 5548 cannot be explained using a standard accretion disc model and that the geometry must be modified to fit the observations.

No coherent explanation is given for how the black holes get there, although Rees points out that any large concentration of mass will eventually evolve towards a black hole. Nevertheless, the time-scale for this is not clear and, in any case, some spectacular events must occur during formation of the black holes.

2.2 STARBURST MODEL

Although the starburst model has been around since before the discovery of QSOs, it had not been considered a viable explanation for AGN because of the presence of high-excitation lines in the spectra of Seyfert galaxies and QSOs.

However, the model was given a new lease of life when Terlevich and Melnick (1985) proposed that the required level of ionisation could be produced by very massive metal-rich stars from which the outer layers had been stripped by stellar winds, resulting in very high photospheric temperatures. These stars they christened "warmers". Variability is then caused by supernova explosions and activity in their remnants, as the stars in the cluster evolve. Terlevich and Melnick (1988) show the spectrum of a "flare" in the Seyfert galaxy,

NGC 5548, that had an emission-line spectrum similar to the extragalactic Type II supernova, SN 1983k, although the continuum slopes were different. Interestingly, Shklovskii had suggested, in his 1960 paper, that the nucleus of M87 should be monitored to look for such supernovae.

Metallicity is an important factor in the evolution of warmers, since high metallicity is needed to provide sufficient line-blanketing in the UV to generate the required radiation-pressure driven winds. This proposition is supported by evidence that late-type galaxies tend to form lower-excitation HII nuclei, while early-type spirals, with higher metallicity, form higher-excitation Seyfert nuclei (Terlevich et al. 1987).

Although the starburst model has not been widely accepted, there is much circumstantial evidence to make it an attractive one. For example, starburst galaxies are known to be associated with galaxy interactions, as are the ultraluminous IRAS galaxies (Scoville and Soifer, 1991) and also nearby QSOs (see, for example, Hutchings et al. 1982, who also note bursts of star formation in the companion galaxies). Although the manner in which jets might form is less clear-cut than in the black hole model, since there is no spinning single central object, the discovery of strong winds in starburst galaxies (Chevalier and Clegg 1985), and of jets from young stellar objects, suggests mechanisms by which the larger-scale jets of AGN might originate. The presence of dust is explained in straightforward fashion, being ejected by the massive stars in the burst as they evolve.

As with the black hole plus accretion disc model, blazars would still require a jet structure with special geometry aligned with the observer.

3. Variability of AGN

What can variability tell us about AGN? Clearly there is information in the light curves themselves which can be related to the nature and perhaps the observed morphology of the objects. Similarly, the light curves might be examined for evolutionary trends with redshift.

In addition, the way in which observations at different wavelengths relate to each other can give clues as to the structure of the source. For example, the reverberation mapping or light echo method (Bahcall et al. 1972; Blandford and McKee 1982) allows something of the structure of the Broad Line Region to be determined. In this, it is assumed that variations in the continuum are followed by variations in the emission lines, according to light-time delays (Fig. 1). Although this only gives a one-dimensional view of the structure, it can be used to put constraints on the size of the Broad Line Region: measurements of this parameter have tended to come out at up to an order of magnitude smaller than predicted by photoionisation models (Penston 1991).

Both black hole and starburst models have difficulty in accounting for the very rapid variability observed in the X-ray region. In the accretion disc model, the speed of these variations means that they cannot be generated globally by reprocessing in the disc and require flares or flashing locally within the disc (Begelman and De Kool 1991). For the starburst model, it has been suggested that the X-ray variations might be caused by winds from the massive young stars. However, in this case, finding an explanation for how the required level of energy is generated presents problems (see the discussion following Begelman and De Kool 1991).

As mentioned above in the case of NGC 5548, the question of whether or not variations at optical wavelengths are delayed with respect to the UV is of particular interest, since the absence of a delay will put severe constraints on the accretion disc model.

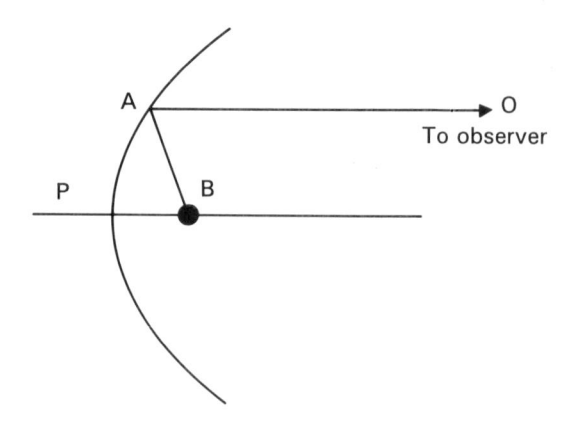

Fig. 1: Reverberation mapping. Changes in ionising continuum at B are reflected in changes of emission line strengths from the ionised region at A. The emission line changes are delayed relative to the continuum by the travel time via BAO compared with that directly from B. The surface of constant delay is a paraboloid, focus A, pole P.

Also of interest is the extent to which infrared variations lag behind the optical. If the infrared component is due to thermal emission from dust, then the emitting region involved is considerably larger than that for a compact non-thermal synchrotron source, i.e. around 10 pc as against 0.1 pc (Bregman, 1991). This should produce an appreciable delay in the infrared variations as compared with the optical/UV, as is indeed found in the case of GQ Comae (Sitko, 1991), supporting the dust model for that particular object. Any delay in the case of the non-thermal model, thought to apply to blazars (Wills, 1991), will be much less.

3.1 MONITORING REQUIREMENTS

The irregularity and wide range in the time-scales of variability in the QSOs make allocation of adequate observing time on conventional general-purpose telescopes difficult. Yet the observations have a fundamentally important role to play, not only in understanding the nature of AGN and associated cosmological problems, but also in the study of peripheral problems, particularly of star formation and galaxy interactions.

There are likely to be many similarities between the observed properties of both starburst and black hole models, since both involve a large mass at the centre of a galaxy and strong UV radiation inducing ionisation in surrounding gas. Ultimately, since the fate of a cluster of massive stars is likely to be a black hole, or something very close to it, there is a question of just when this will happen and whether or not both types of object can be observed at the present time. However, if one starts with a black hole, then the accretion disc model does not explain how it got there; the starburst model gives a reasonable evolutionary track to such a situation.

Although monitoring at different wavelengths is important in establishing models of the Broad Line Region, limited wavelength observations will still give important clues and restrictions on models. For example, measurements in the infrared and optical will allow a distinction to be drawn between whether the infrared emission is due to warm dust or to non-thermal processes.

Since, in the starburst model, variability (blazars excepted) is ascribed directly to supernovae and activity in their remnants, the light curves should show transient phenomena over periods of weeks, with slower variability over periods of years due to the remnants; this type of behaviour is clearly shown in the light curve of the quasar 1510-089 and perhaps also 3C 120 (Smith et al. 1991). It is also likely that, for more luminous AGN with overlapping supernova events, the relative variation in magnitude will reduce because of a "root-n" effect, although the noise flux will still increase in absolute terms.

Supernovae will not always be visible, depending upon the evolutionary stage of the AGN, which will thus provide a further test for the starburst model.

It is important to obtain spectrophotometric data as well as photometry in order to permit reverberation calculations to be carried out and, similarly, multi-wavelength campaigns, like that carried out on NGC 5548, will be crucial. But it is also essential to obtain long runs of well-sampled optical photometry, and robotic telescopes are the only realistic prospect of obtaining such systematic and well-calibrated data.

4. Scientific returns

An optical monitoring campaign, carried out on a robotic telescope, should provide information on many important questions, especially when combined with data at other wavelengths. Are the principal engines of AGN black holes plus accretion discs or starburst, or are both observed from our present cosmological vantage point? In how many cases does the supernova-plus-remnant fit the observed light curve and does the proportion of such cases match the predictions of the starburst model? Is the Broad Line Region always smaller than predicted by the photoionisation models and can this effect be accommodated into the black hole model reliably? What part do mergers play in galaxy evolution and does metallicity give clues to this? How far does the relation between the type of AGN and metallicity go and can this be related to the morphology of interacting pairs? Indeed, are spiral arms themselves formed as a result of merger activity? Is the occurrence of QSOs in elliptical galaxies indicative of a merger or leftovers from a recent merger?

Whatever the answers to these questions, the result of extensive AGN monitoring campaigns is likely to have a profound effect on our understanding of galaxies and their evolution.

References:

Bahcall, J.N., Kozlovsky, B.-Z. and Salpeter, E.E., 1972, Astrophys. J., **171**, 467.
Begelman, M.C. and De Kool, M., 1991, in: *Variability of active galactic nuclei*, eds. R.H. Miller and P.J. Wiita, Georgia State Univ. Conf., p. 198.
Blandford, R.D. and McKee, C.F., 1982, *Astrophys. J.*, **255**, 419.
Bregman, J.N., 1991, in: *Variability of active galactic nuclei*, eds. R.H. Miller and P.J. Wiita, Georgia State Univ. Conf., p. 1.
Chevalier, R.A. and Clegg, A.W., 1985, *Nature*, **317**, 44.

Clavel, J., 1991, in: *Variability of active galactic nuclei*, eds. R.H. Miller and P.J. Wiita, Georgia State Univ. Conf., p. 301.
Done, C., Ward, M.J., Fabian, A.C., Kunieda, H., Tsuruta, S., Lawrence, A., Smith, M.G. and Wamsteker, W., 1990, Mon. Not. R. astr. Soc, **243**, 713.
Hutchings, J.B., Campbell, B. and Crampton, D., 1982, Astrophys. J. (Letters), **261**, L23.
Molendi, S., Maraschi, L. and Stella, L., 1992, Mon. Not. R. astr. Soc, **255**, 27.
Penston, M.V., 1991, in: *Variability of active galactic nuclei*, eds. R.H. Miller and P.J. Wiita, Georgia State Univ. Conf., p. 343.
Rees, M.J., 1984, in: *Annual Review of Astronomy and Astrophysics*, ed. G. Burbridge, **22**, p. 471.
Scoville, N. and Soifer, B.T., 1991, in: *Massive Stars in Starbursts*, eds. C. Leitherer, N.R. Walborn, T.M. Heckman and C.A. Norman, *STSI Symp. Series,* **5**, p. 233.
Shklovskii, I.S., 1960, Soviet Astr., **4**, 885.
Sitko, M.L., 1991, in: *Variability of active galactic nuclei*, eds. R.H. Miller and P.J. Wiita, Georgia State Univ. Conf., p. 104.
Smith, A.G., Nair, A.D. and Clements, S.D., 1991, in: *Variability of active galactic nuclei*, eds. R.H. Miller and P.J. Wiita, Georgia State Univ. Conf., p. 52.
Terlevich, R. and Melnick, J., 1985, Mon. Not. R. astr. Soc, **213**, 841.
Terlevich, R. and Melnick, J., 1988, Nature, **333**, 239.
Terlevich, R., Melnick, J. and Moles, M., 1987, in: *IAU Symposium 121: Observational Evidence of Activity in Galaxies*, eds. E.Ye. Khachikian, K.J. Fricke and J. Melnick, p. 499.
Turner, T.J., Kunieda, H., Mushotzky, R.F. and Awaki, H., 1991, in: *Variability of active galactic nuclei*, eds. R.H. Miller and P.J. Wiita, Georgia State Univ. Conf., p. 260.
Wills, B.J., 1991, in: *Variability of active galactic nuclei*, eds. R.H. Miller and P.J. Wiita, Georgia State Univ. Conf., p. 87.

The concept of an APT network as a driver for a metrological reform in astronomical photometry

C. Sterken[1]

University of Brussels (VUB), Pleinlaan 2, 1050 Brussels, Belgium.

Abstract

The basic problem of networks of automatic photometric telescopes (APT) capable of yielding data with millimagnitude precision is that of homogenisation of the data. Homogenisation can be achieved only when the measurements are made in photometric systems that are truly transformable, i.e. when the actual system of astronomical photometric measurements is thoroughly reformed. This paper assesses some of the problems to be expected if such a reform comes true.

1. Introduction

Reviewing the actual status of photometry and photometric systems is no simple matter. Such a review not only must assess the activity of the system in terms of numbers of users, the rate of collection of data, and the geographical spread of the sites where the system is implemented, it must also assess the ever-expanding field of astronomy that must be covered by any system that wishes to survive. One must, in addition, estimate the impact of the new imaging detectors that are becoming more and more efficient and reliable. Finally, one must judge the increasing need of a basic requirement that is becoming more and more a principal necessity: the applicability of the photometric system in a global network, i.e. the stringent requirement of homogeneity.

The first point simply means that photometric systems are increasingly used for applications for which the system initially was never designed (for example, the *uvby* system is now wildly applied to the study of exotic stars, such as WR and symbiotic stars). True, these are often studies of variability, and may make sense only when carried out with great care and with true consideration for the basic assumptions that inevitably are being violated. It is clear that the widening of the field of applicability is not necessarily a sign of increasing importance of the system, but that it is rather an indication of accessibility, i.e. the fact that a specific photometric system is applied is rather the consequence of its local availability (especially at observatories with few visitors) than that it is the consequence of real necessity.

It is no exaggeration to say that many observers do not even realise that they are trespassing on the domain of validity of the specific system they apply. Such a situation can only enhance the exposure of that system's faults, and hence can only contribute to extinction of that system itself.

The second point, the impact of new detectors on the death of old systems and on the birth of new ones stands clear when one looks at Fig. 1 of Sterken (1992), which is a graphical interpretation of the table of ancestry for some existing photometric systems (only the well-known systems are indicated). The four main branches (visual, photographic, photomultiplier tube and CCD) are detector branches, from which the individual systems have grown[2]. Note that this representation is an idealised situation in the sense that it depicts

---

[1] Belgian Fund for Scientific Research (NFWO).

the view as seen by the designer of each system. It does not show the reality in which several "pure" systems are accompanied by "clones" which are incompatible with each other and also with the original system they came from. The cloning is a direct consequence of both the openness of the system, and of the progress in detector development.

Finally, it is perhaps the third element, viz. the establishment of worldwide networks, and the ever-increasing need to compare, combine and merge data coming from one observer's site with results provided at another scientist's telescope that will shape the future. In this process, homogeneity is a vital factor.

It is a common misconception of photometrists that adequateness of the hardware (i.e. proper filters, a correct detector, and a high-tech photometer) are a sufficient condition to produce state-of-the-art results. Besides proper hardware, one must have an appropriate photometric system.

The introduction of intermediate bandwidth systems, such as the Strömgren *uvby* system, has not fulfilled the promise of elimination of disturbing bandwidth effects. Though extinction effects of this origin are proportional to the square of the bandwidth,[3] the problem with reddened or exotic-type stars simply remains, as was shown by Manfroid and Sterken (1987), who based their conclusions on carefully conducted *uvby* measurements collected at a good site using a well-designed photometer and obtained with proper consideration for standard-star measurements. Similar discrepancies were also found in other photometric systems (see Blanco 1957, Mandwewala 1976, etc.). It is stressed that the application of differential photometry is in no way in a position to remove those effects; on the contrary, differentiation may even enhance such anomalies. The trouble simply is that *the transformation problems are part of the system*, i.e. that they are an unavoidable consequence of the basic specifications built into the original design of the photometric system itself.

The problem, as every photometrist knows, lies in the fact that the transformation equations involve the derivatives of the spectral irradiance of the observed object, modified by the passage through the atmosphere and through interstellar space (usually, only the first and—occasionally—the second derivatives are effectively used). In practice, the reduction procedures assume that these first derivatives can be approximated by using the colour indices of the object in question. This approximation does work for most stars that are unreddened and that have spectral energy distributions which do not violate the Taylor series expansions. Still, Young (1992) demonstrates that the classical series expansion in King's theory (King 1952) needs to be carried out to at least fourth order if millimagnitude accuracy is to be achieved.

However, for emission-line stars, symbiotic and other peculiar stars, and for highly - reddened objects and even for multiple stars with components belonging to different colour-index classes, the method breaks down, and even "normal" stars yield colour indices that only coarsely represent the first derivatives, as has been emphasised already by Young (1974). Wider agreement among observers will be achieved when such improved and refined application of known principles is commonly adopted.

The ultimate problem simply is the fact that every photometric system (though some systems much more than others) estimates the derivatives incorrectly because they do not have sufficiently overlapping passbands (some systems have no overlapping passbands at all). Young's remedy (Young 1974, 1992) is to define a well-sampled—thus transformable-

---

[2] Again, this is largely a matter of availability, without regard to photometric considerations.

[3] In the colour transformation, they are inversely proportional to bandwidth; see Young (1992).

system that is compatible with existing systems like UBV and $UBV\,B_1B_2V_1G$, and where the bands are limited in a way which is totally independent of the colour of the star, and where every local instrumental configuration very well duplicates the original standard system. Such a system will at the same time yield better estimates for interstellar and atmospheric extinction corrections. More generally, Young (1993) demonstrates that unavoidable manufacturing variations in passband shapes will not produce disturbing higher-order effects in transformations from one system to another, if the passbands of one filter set are linear combinations of the passbands of the other and if the transformations involve intensities and linear combinations of intensities, not magnitudes and classical colour indices. Only then will a unique, and accurate, transformation be applicable to all observed objects, from far-away galaxies to even heavily obscured multiple systems with spotted components of whatever type.

2. Historical aspects of metrological reform

In fact, what Young proposes, is a true metrologic reform in the field of astronomical photometry: *replace traditional photometric metrology*[4] *by a system that is homogeneous, commensurable, and has a non-variable measure*, i.e. has basic standards of constant character.

History may perhaps help us to assess the problems to be expected if Young's reform ever comes true. Let us make a short review of what happened two centuries ago when the French Revolution brought the metric reform of weights and measures in France and in the countries ruled by France.

In those days, every town had its own standard. A given unit of length recognised in Paris, for example, was about 4 per cent longer than that in Bordeaux, 2 per cent longer than that in Marseille, and 2 per cent shorter than that in Lille (Klein 1974). This does not necessarily mean that the ancient system of measurement was defective. Old units of surface, for example, were related not only to surface, but mainly to the associated intensity of agronomic labour, and those measures henceforth were much more additive than are any contemporaneous measures of surface. Also, locally at least, some surface units were proved constant over time intervals of several centuries.

Why, then, was a reform so needed? The reform really took place because there was an urgent need for constant measures, based on a commonly adopted standard unit, due to the rapidly growing *networks of international collaboration* in commerce and trade, carried by technological advances in all fields. In addition, there was the will and support of governmental bodies, who in turn were instigated by scholars and scientists, not seldom driven by progress in mathematical sciences.

Kula (1970) describes three phases in the evolution of measurement. The first phase is characterised by a purely anthropomorphic conception of measurement, where man measures the world with his own body (or with a body) as standard (i.e. the use of foot, inch, pace...), and this approach is one of all times and cultures. Later, the spectrum of measures is extended beyond the proper size of the body, and body-related measures (associated with man's activity and production) are introduced (e.g., stone's throw, bowshot); every new technological development brought with it a growth of related measuring systems,

---

[4] In astronomy.

and these early weights and measures proliferated in urban and rural areas. The last stage is the removal of the human link, and the introduction of a purely abstract system of measure, with a unit based on astronomical measurements. The subsequent phases correspond to increasing degrees of accuracy, and to increasing facilitation of collaboration. In other words, there is a convergence of increasing precision with (expanding) collaboration and unification.

The metric reform was not an easy transformation, and, in fact, the metrologic revolution took more than half a century to become effective. This was mainly because of the presence of obstacles of psychological order, viz. passive and active opposition, and also plain inertia.

Though a strong will for reform was present, the desire was not so obviously shared by the merchants. After all, it was thought the old system had not severely hampered the development of international trade. Two decades after its installation, the new metric system was taught in schools, and used only in public administration and in some circles of upper commerce. But the old weights and measures that were to have been destroyed remained in use all over the country in daily trade, partly due to the ignorance of the people, and mainly because of entrenched habits. In 1812, a step backward was taken by opening up the possibility of the adaptation of metric measures for retail trade, allowing non-decimal fractions (1/16, 1/8...) of metric units, and even replacement of older units by metric approximations (as the inch is now defined as 2.54 cm exactly). These measures were called *mesures usuelles*. Thus, a new and useless system had been introduced: it looked like the old one, but was not sufficiently identical to the old one.

It was soon recognised that the 1812 concession of metric accommodations to the older feudal units had caused more negative than positive effects: the measure not only strengthened passive and active resistance, it also forced the Administration to control two different systems: one for mass use, one for trade at higher levels. Therefore, after a short time, it was proposed to forbid any other system, and allow only the official one (Bigourdan 1901), and in 1837 the Minister of Trade introduced a project of law[5] that asked for the reinstallation of the pure metric system.

What had been the elements of opposition against the new system? First of all, there was the force of established custom. Second, the decision, and its enforcement, had a political colour. In the *Moniteur* of 25 June 1798, an analysis of these elements was presented, and this analysis revealed a remarkable additional element. The report states that 50 kilograms of platinum had been purified in order to make the prime standards, and clear instructions for exact duplication (i.e. for the production of secondary standards) had been provided. Consequently, it was stated that the delivery of a set of weight and length standards to the merchants of the whole republic will cost much more than what the public treasury can provide, and that hence the merchants themselves should pay for these standards. Finally, it was very clearly stated that, *if the use of the system did not spread as fast as one had hoped, this was not due to a defect of the system itself, but that this was eventually due to shortage of funds, because standards could not be fabricated in a sufficient number* to allow for the needs of the whole (French) society.

---

[5] 7 messidor an VI in the reformed Republican calendar, a reform that was eventually abandoned.

3. The reform of photometry seen against the historical background

The evolution of photometric measurement, of course, reflects the three stages of evolution of measurement brought up in the preceding section. The tree of genealogy cited above expresses this evolution: subsequent changes of detector (eye, photographic emulsion, and photomultiplier) have produced data of higher levels of precision that progressively move further and further from an anthropocentric basis (e.g., from m_v to m_{pv} to V). Abstraction, and the introduction of a purely artificial basis, clearly is the only way to progress.

There are several photometric systems that, locally at least, were proved constant over time intervals of several decades, and this stability was greatest for those local systems that were able to "defend" well their measures, whereas other, more "rural" systems underwent stronger variations. *UBV* and *uvby*, respectively, are the wide-band and intermediate-band leaders in the class of open systems (for a discussion of the concepts of "open" and "closed" systems, see Rufener 1985). There are numerous closed systems with different degree of rigidity concerning the aspect of closedness. The Geneva $UBV\ B_1B_2V_1G$ system, which certainly is the world's most closed system, has accomplished the remarkable fact of combining high accuracy, internal homogeneity, and broad coverage both in number of measurements and in the diversity of objects studied (stars of all spectral types and luminosity classes, supernovae, QSOs, galaxies, solar-system bodies, etc.).

There is, though, a strong need for a reform, as has been stressed in the introduction, and in references quoted therein. The plea for a structural reform is made by scientists, and is supported by "governmental" bodies, such as the IAU. The pitfalls that must be avoided stand out clearly. One should avoid the introduction of any newly-bred systems that both seem to fit the old traditions (existing photometric systems) and the new detectors. In other words avoid, by all means, the creation of any *"mesures usuelles"*. And, although *by design* the new system is so that *exact* reproduction of filters is *not* required, newly designed filters must be made in very large batches in order to minimise the unavoidable internal inhomogeneities and mutual differences, and those filters should be distributed to observers worldwide, free of charge. The last point is especially true for small observatories, or institutes in countries with modest science budgets.

One might also argue that there is to be expected a strong opposition of supporters of closed systems, who may want to protect, or even impose, their own heritage of invariable, ancestral measures. History, however, has shown that such was never the case. It is exactly the rich townships, that had the most stable and geographically most widely spread metric tradition, that were the first to acquire the (costly) standards and to accept the reformed system. The large society of very open, variant, local systems was much more hesitant to follow, and the strong will to reform was rather not shared by the merchants, i.e. the common users.

Supporters of a reform in photometry might ask whether one should not abolish the outdated logarithmic magnitude scale, and replace it by a linear intensity scale. Whereas the use of intensity in the reduction calculations is one necessary condition to allow for perfectly linear transformations, there is no objection against converting the results back to usual magnitudes. A resolution to abandon the traditional magnitude scale may, in fact, turn out to be more short-lived than was the ephemeral Revolutionary calendar.

4. Conclusions

The existence of a just system of measurements has always been a criterion of civilisation. Modern astronomical photometry still lacks a coherent system that will allow "just measures". The largest problems in photometry are of a structural character and of methodological nature, rather than of technological origin or due to lack of existing observing facilities. *These problems will not be solved by the introduction of networks of automatic telescopes*, but, on the contrary, these problems will become larger and larger. Together with the development of robotic networks, one should give proper consideration to some of the suggestions mentioned in the previous section, specifically concerning the elaboration and implementation of a new photometric system. Such a transformation, of course, will encounter opposition. One should remember, however, that the introduction of the metric system in France two centuries ago, took several decades to take hold. Though the reform of photometry is a very much less ambitious idea, and although scientists have become very cosmopolitan in the deployment of networking activities, one may only regret that several developments in automatic photometry still seem to occur on very narrow bases, confined to single nations, even to single institutes or research teams. The photometric community, has an interest to search for funds for undertaking such a project in a truly international environment (see also the chapter by Crawford in Part III of this volume).

References:

Bigourdan, G., 1901, *Le système métrique des poids et des mesures*, Gauthier-Villars, Paris.
Blanco, V.M., 1957, *Astrophys. J.*, **125**, 209.
King, I.R., 1952, *Astron. J.*, **57**, 253.
Klein, H.A., 1974, *The science of measurement*, Dover, New York.
Kula, W., 1970, *Miary i ludzie*, translated 1984, *Les mesures et les hommes*, Editions de la Maison des sciences de l'homme, Paris.
Mandwewala, N.J. 1976, *Arch. Geneve*, **29**, 119.
Manfroid, J. and Sterken, C., 1987, *Astron. Astrophys.* Suppl. Ser, **71**, 539.
Rufener, F., 1985, in: *Calibration of Fundamental Stellar Quantities*, p. 253, eds. D.S. Hayes, L.E. Pasinetti and A.G. Davis Philip, Reidel, Dordrecht.
Sterken, C., 1992, On the future of existing photometric systems, *Vistas in Astronomy*, vol. 35, p. 139.
Young, A.T., 1974, in: *Methods of Experimental Physics*, **12A**, ed. Carleton, N., Chapters 1, 2 and 3.
Young, A.T., 1992, Astron. Astrophys., **257**, 326.
Young, A.T., 1993, in: *Stellar Photometry, IAU Coll. No. 136*, eds. Butler, C.J. and Elliott, I., Cambridge University Press, p. 80.

How accurate are photometric standards?

T. Oja

Astronomical Observatory, Box 515, S-751 20 Uppsala, Sweden.

Abstract

Two lists of equatorial UBVRI standards have been inter-compared and show unexpectedly large systematic differences.

1. Introduction

Practically all photometric measurements have to be related to some kind of standards in order to be useful. This is without exception the case when magnitudes and colours are measured with instrumentation not used when defining the system, i.e. in all measurements in the UBVRI system for instance. Generally the quantities in the standard system are expressed as (more or less) linear functions of quantities in the instrumental system. The derivation of these functions (and sometimes of atmospheric extinction) presupposes the existence and availability of a sufficient number of standard stars at suitable zenith distances, the magnitudes and colours of which are known with high precision.

The UBV system was originally defined by a list of standard stars (Johnson and Morgan 1953, Johnson 1954). This list is dominated by bright stars at northern declination, so it cannot be used directly always and everywhere. It was transferred to the E regions at Dec. $-45°$ by Cousins and Stoy (1962), and two sets of equatorial standards, also including R and I magnitudes, have subsequently been established (Landolt 1983, Wall et al. 1989). The equatorial standards contain stars in a wide magnitude range and some standards are thus always available also with big telescopes from all observatories not located too far from the equator. These stars are also frequently used for the calibration of CCD photometry.

2. Inter-comparison of data sets

The two sets of equatorial UBVRI standard stars have a considerable number of objects in common, and it was considered to be of some interest to inter-compare them. The result of the comparison is shown in Fig. 1. Disregarding a few late-type dwarf stars and known variables the conclusion is:

(i) V-R and R-I are in excellent agreement except perhaps for very red stars.
(ii) There are small but significant linear colour equations in V and U-B.
(iii) There is a considerable, non-linear colour equation in B-V.
(iv) The dispersion around the mean relations is small, ± 0.005 for V, B-V, and V-R, ± 0.007 for R-I and ± 0.011 for U-B.

The internal accuracy of both series of data is obviously very high and renders them well suited for their intended purpose as standards. The colour equations, however, are alarming, especially the one in B-V.

3. Conclusions

At this stage it is not possible to tell which set of standards is more correct than the other one, or in what way "best" or "correct" values can be derived from existing data (except perhaps by assuming *a priori* that a systematic error of the order of magnitude of ±0.02 in accurate B-V photometry is not possible and would have been revealed immediately,

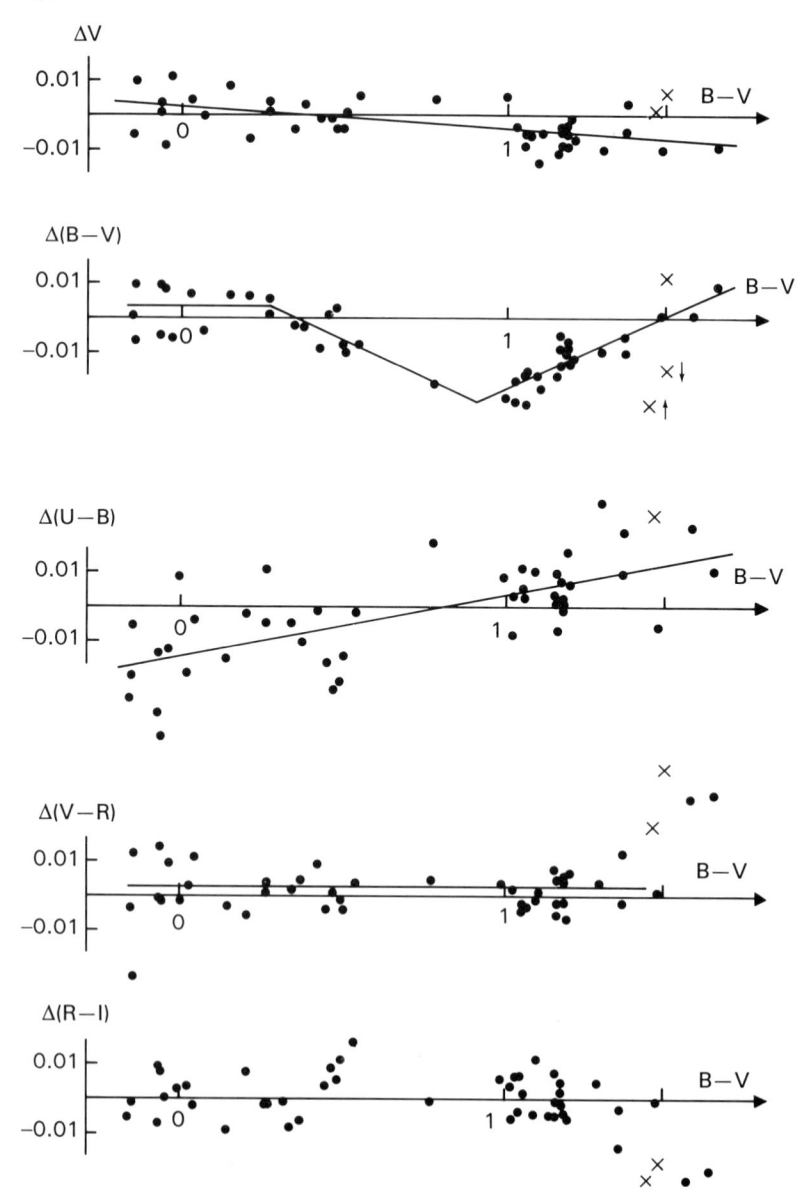

Fig. 1: The differences between the two sets of data as functions of B-V (in the sense Landolt minus Wall et al.); crosses denote M dwarfs.

so the mean system—leaving a systematic error of about ±0.01—is certainly better than either set of data alone?). For neither list of standards has the retrieved photometry of the primary standards been published.

New independent observations are planned in collaboration with Dr. D. Jones at the Royal Greenwich Observatory in order to solve the controversy. Until further results are at hand one has to keep in mind that the lists of secondary UBVRI standards have to be used with caution.

It should finally be stressed that it is of great importance that the resulting values of the standard stars (reduced in exactly the same way as the programme stars) are published together with all photometric lists; had this been the case for the equatorial standards, it would have been easy to trace the cause of the discrepancy.

References:

Cousins, A.W.J. and Stoy, R.H., 1962, *R. Obs. Bull.*, 49.
Johnson, H.L., 1954, *Ann. d'Astrophys*, **18**, 292.
Johnson, H.L. and Morgan, W.W., 1953, *Astrophys. J.*, **117**, 313.
Landolt, A.U., 1983, *Astron. J.*, **88**, 439.
Wall, J.V., Laing, R.A., Argyle, R.W. and Wallis, R.E., 1989, *ING La Palma User Manual No. XII*.

Editor's note:

The result of the re-observation of a number of equatorial standards, together with primary UBV standards has been published recently in Oja, T., 1994, *Astron. Astrophys.*, **286**, 1006.

Towards robotic IR observatories: improved IR passbands

E.F. Milone,[1] C.R. Stagg,[1] and A.T. Young[1,2]

[1]Department of Physics and Astronomy, University of Calgary,
2500 University Drive, N. W., Calgary, Alberta, Canada T2N 1N4.
[2]Astronomy Department, San Diego State University, San Diego, CA 92182, USA,
and European Southern Observatory, Garching bei München, Germany.

Abstract

An improved set of infrared passbands promises to open up a new vista for infrared astronomy by permitting observations at lower altitudes, under less favourable circumstances, than previously possible with the traditional and most of the newer infrared passbands. While a full description of the system will be given elsewhere, we emphasize the aspects of the improvements which may lead to fully automated infrared observatories.

1. Introduction

Non-thermal infrared photometry is capable of higher precision than is photometry in the visible spectrum. This situation is ironic because, currently, atmospheric extinction correction is more difficult in the infrared. Over observable ranges of airmass, the extinction appears to be small; however, magnitudes cannot safely be extrapolated above the atmosphere because strong and variable water-vapor absorption intrudes into almost all existing infrared passbands, except those few specifically designed to eliminate it. Because water vapor varies from place to place, from day to day, and even from hour to hour, infrared astronomy has lower precision and accuracy than would otherwise be possible. The situation is reviewed in Milone (1989).

It is indeed a tribute to the ingenuity and hard work of infrared astronomers that transformations have been possible at all. The work has required painstaking attention to the conditions of observation and the standardization of filters and equipment, and the restriction of observing conditions, and observing sites. Bessell and Brett (1989) summarize the careful work done to standardize infrared photometry as it is currently practiced. Recognizing the limitations of the current situation, a Working Group of Commission 25 on Infrared Extinction and Standardization has been implementing the suggestions made at a joint meeting of Commissions 25 and 9 at the Baltimore IAU General Assembly on just this topic.

Obviously, the situation can be improved by reducing, as far as possible, the response of the instrument to wavelengths where water absorbs strongly. Naturally, a compromise must be struck between the need to minimize the influence of water vapor, and the desire to maximize the measurable signal. The dilemma is not as severe as one might suppose, because not much light is transmitted where the water bands are strong; so by giving up the atmosphere-defined portion of the bandpass, the extinction problem can be reduced by a large factor, and the transformability of the passbands improved. Moreover, the relative insensitivity of cleaner passbands to water vapor means that infrared photometry will be possible from sites other than the highest and driest, to which it has been confined until now.

Here we present the simulated extinction curves for a typical existing passband and for our suggested passband, which offers a real prospect of millimagnitude precision for infrared photometry.

Since there are other obstacles to be overcome, such as the need for cooling without physical cryogen, the process described here is a necessary but not sufficient step to fully automated infrared astronomy.

2. Infrared versus optical extinction

In broadband photometry, a plot of magnitude versus airmass is more or less curved, because the wavelengths with the largest monochromatic extinction coefficients are removed at small airmasses from the stellar spectrum, leaving the wavelengths with smaller extinction coefficients at larger airmasses. Thus, the slope of the line continuously decreases with increasing airmass. This effect, which is strong in the infrared (Manduca and Bell 1979, Volk et al. 1989), was discovered by J. D. Forbes (1842) and is named after him.

In the visible (or, in the parlance of IR astronomers, the optical), the monotonic run of extinction with wavelength produces a correlation between extinction coefficient and stellar color index. Here the Forbes effect can be thought of as a progressive decrease in extinction due to the increasing atmospheric reddening at large airmasses (King 1952, Young 1974, 1988). This is not the case in the infrared.

In both the visible and the infrared, the Forbes effect can be regarded as a curve-of-growth effect involving the progressive alteration of the transmitted stellar spectrum. As is well known, the effect of atmospheric reddening in B can be removed quite accurately, using a small linear color term. Typically, the B extinction coefficient varies by about 0.03 mag/airmass per unit of (B-V) color index. Moreover, the Chappuis bands of ozone almost fully compensate for the wavelength dependence of Rayleigh scattering across the V band, so that the extinction is nearly constant across V, making a color term unnecessary.

However, whereas the effect in the visible is small, and can be described by a single parameter (essentially the product of the color coefficient in the extinction with the reddening per unit airmass), the curved infrared extinction line requires at least three parameters. In this study, we used Young's (1989) parameterization of the curve.

Because the correlation with color is much weaker in the infrared, we cannot expect to be able to remove such effects as fully there. In the infrared, the atmospheric transformation errors may remain nearly as large as differences between instrumental and standard systems in the visible before color-term transformation, so large systematic errors between stars with different spectra can be expected.

Using an approach in which the stellar irradiance and instrumental response functions are considered vectors in Hilbert space (Young 1993, 1994, Young et al. 1993, 1994), we can use the angle through which the atmosphere rotates the passband's Hilbert-space vector as a measure of the Forbes effect. If this angle θ is small—say a few degrees at most—we can expect minimal difficulty. If $\theta > 20°$, we can expect substantial problems, as our simulations have demonstrated. We have tried to choose band profiles that keep the atmospheric effect down to a few degrees, but that also provide maximum transformability while retaining a respectable throughput.

3. Extinction curve simulations

We have used the AFGL program MODTRAN to tabulate typical atmospheric transmission values t(v,M) at 1 cm^{-1} intervals, and combined these with model-atmosphere fluxes I$_v$(λ) kindly supplied by Kurucz. To reduce the size of the parameter space to explore, we omitted aerosol extinction entirely, and concentrated on models with different water contents, because aerosol extinction varies only slightly across IR bandpasses but water-vapor absorption is important, varies rapidly with wavelength, and depends strongly on site and season. For each model, we computed the transmission, t, at every half airmass between 1.0 and 3.0.

To minimize computation, we selected a subset of the stellar-atmosphere models, with log g = 0.0 and 4.0 and ten temperatures from 3,500 K to 35,000 K, plus the solar and Vega models, from Kurucz's extensive tabulation. These fluxes and a number of published as well as unpublished response functions, R(v), were interpolated to 1 cm^{-1} resolution. The three functions were multiplied together, and the products summed to approximate the wavelength integral

$$L(M) = \int_0^\infty I(v)\, t(v,M)\, R(v)\, dv \qquad (1)$$

which is the (simulated) measured signal at M airmasses. These signals were converted to magnitudes, in preparation for plotting extinction curves and numerical analysis.

To model the extinction curves, the rational expression used by Young (1989) was first converted to a more practical form. We rewrite his Equation (3) as

$$m = \frac{a + bM + cM^2}{1 + dM} \qquad (2)$$

where a represents the extra-atmospheric magnitude. When $M \ll 1/d$, the initial (and not directly observed) slope at zero airmass is $(b - ad)$. When $M \gg 1/d$, the asymptotic line has the equation

$$m = (b/d) + (c/d)M \qquad (3)$$

Since we expect the extrapolated intercept b/d to be only a few tenths of a magnitude larger than a, it is reasonable to set

$$b/d = a + b'$$

or

$$b = d(a + b'). \qquad (4)$$

where b' depends on the atmospheric transmission, with $1/d$ approximating the airmass at the "corner" of the curve. Note that the vertical asymptote of Equation (2) is at $M = -1/d$; in fact, $1/d$ corresponds exactly to M_0 in Young's (1989) Equation (3).

In the large-airmass limit, the slope approaches the smallest (monochromatic) extinction coefficient within the passband, so we expect c/d to be independent of the stellar spectrum. But b' and d depend on the relative weighting of different wavelengths by the stellar spectral irradiance, and thus should depend on stellar properties; e.g., if the molecular absorption is stronger in the blue wing of the passband than the red one, the Forbes effect will be stronger for blue stars than for red ones. Then b' and d will be larger for blue stars. We have tried fitting linear color dependences for these latter coefficients, but those results are not given here.

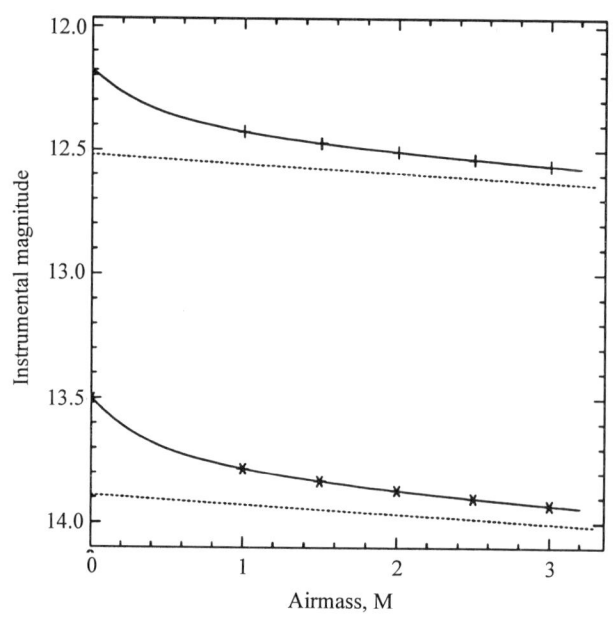

Fig. 1: Extinction curve for Johnson's J band at Mauna Kea. The lower curve is for the model atmosphere with 3500 K and $\log g = 0$; the upper curve is for $T = 5250$ K.

We have done simulations for both the original Johnson (1965) passbands, and for cleaner band profiles that will be described elsewhere (Young et al. 1994). The program *GaussFit* (Jefferys et al. 1988) was used to fit Equation (2) to the simulated observational data which extend from 0.0 to 3.0 airmasses. Typical residuals are a millimagnitude or less; the maximum residuals do not exceed 0.01 mag.

Fig. 1 shows extinction curves for Johnson's (1965) J passbands at a high-altitude site for the tropical atmosphere model available in MODTRAN, typical of conditions at Mauna Kea. The quantity d is typically about 2; the corner occurs near $M = 0.5$.

Fig. 2 shows similar extinction curves for a relatively low-altitude observatory (1 km above sea level), using the mid-latitude summer model atmosphere from MODTRAN. The non-linearity of the extinction curves is much worse; but again the damage is nearly all done between 0 and 1 airmass, and so is nearly invisible to the user who tries to fit a straight extinction line to observations. Comparison of Fig. 2 with Fig. 1 shows that we can expect serious extrapolation errors that vary with atmospheric water content, if the original Johnson passbands are used.

```
          12.0

          12.5
Instrumental magnitude
          13.0


          13.5


          14.0

              0        1        2        3
                        Airmass, M
```

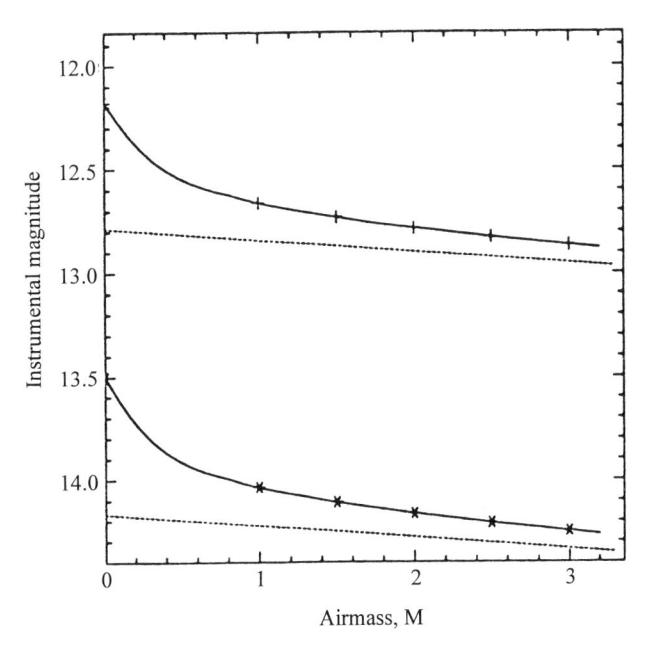

Fig. 2: Extinction curve for Johnson's J band at 1 km in summer. Compare to Fig. 1.

A particular problem with these curved extinction lines is that the extrapolation error of a straight line fitted to the part of the curve at accessible airmasses varies strongly with the water content of the atmosphere, and hence varies from site to site and from night to night. The photometer appears to have an unstable zero point that varies with the water. This means that one cannot use the powerful assumption of a stable instrumental zero in using data from multiple nights (Young 1974, Manfroid and Heck 1984, Schwarzenberg-Czerny 1991) to solve for the extinction.

Finally, Fig. 3 shows extinction curves calculated for a cleaner band profile. Even for the relatively humid summer atmosphere at 1 km, the extinction line is nearly straight. For Mauna Kea, it is also straight. Extrapolation errors are very much smaller here, and we should be able to use multi-night solutions to improve the measurement and removal of extinction. These figures indicate the marked improvement that is possible if one avoids the strong water-vapor absorptions at the sides of the Johnson passbands.

Because Johnson's original J band was nearly centered on the 1.12-μm water band, we have less than half of the original J flux in the cleaned-up J band; the instrumental magnitudes are about 1 mag fainter in Fig. 3 than in Fig. 2. However, the window centered at 1.03 μm is even clearer than the J window at 1.2 μm. By adding a filter at the shorter wavelength, we can get back most of the lost photons, with an excellent straight extinction curve (Fig. 4). In addition, we have gained a useful color index. Thus, the information content of observations made with the two clean bands in place of Johnson's original J should be appreciably greater, even if random errors are a little larger for the faintest stars.

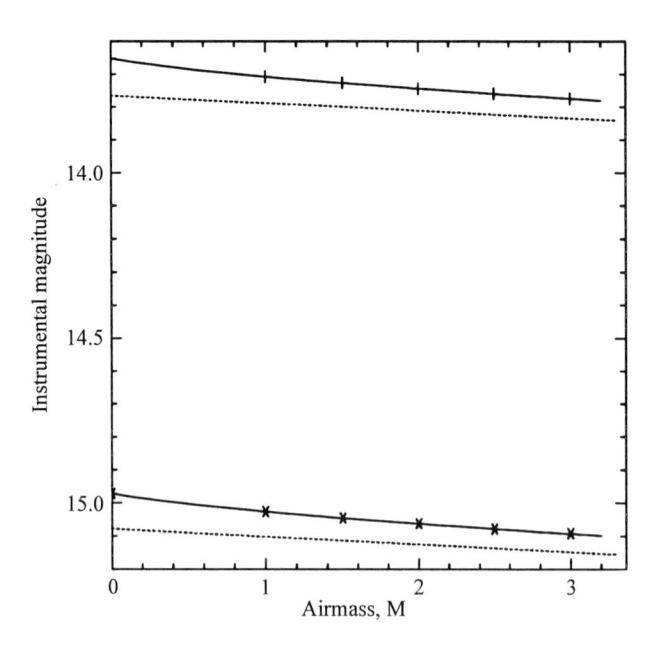

Fig. 3: Extinction curve for improved J band at 1 km in summer. Compare to Fig. 2.

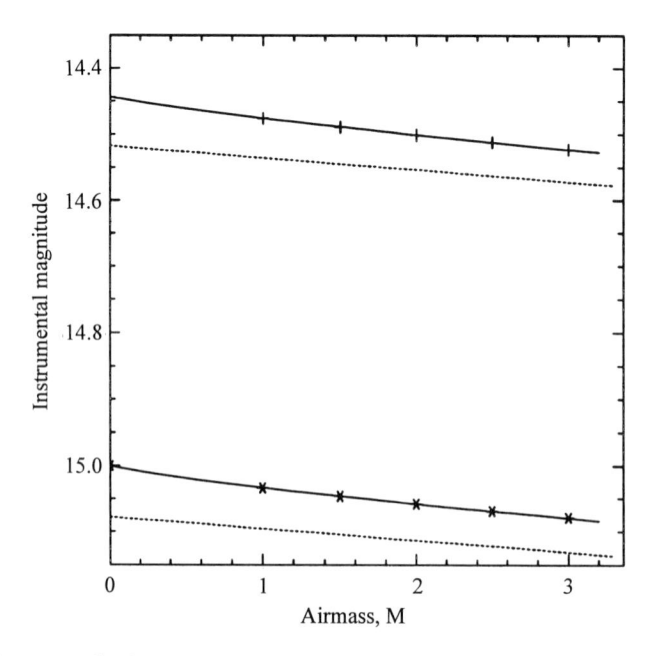

Fig. 4: Extinction curve for improved 1.03-μm band at 1 km in summer.

4. Other considerations for IR robotic observatories

In many telescopes the background emission is dominated by the telescope. Hodapp et al. (1992) cite a contribution from the telescope of 14.5 mag/(arcsec)^2 in Wainscoat and Cowie's (1992) K'-band, and 14.9 mag/(arcsec)^2 from the atmosphere, on a particular night. Unfortunately, the Wainscoat—Cowie filter extends deeply into the water on the short-wavelength side, so it has nearly as bad an extinction curve as the original Johnson K (Young et al. 1993). The use of superior filters can therefore help, but attention to clean telescope and photometer design is still required for success in the thermal infrared. Automated mirror cleaning would improve the performance of IR robotic telescopes.

All IR detectors require cold operating temperatures: LN$_2$ or colder. At present non-liquid cryogen coolers are very expensive, and although pumping on the cryogen container can reduce temperatures and minimize subsequent evaporation rate, replenishment must still occur in one to two weeks (sooner if there has been a power interruption). Continuous, automated cooling facilities must be developed for unattended robotic IR observatories.

Acknowledgements

We thank Bob Kurucz for providing his model-atmosphere fluxes; Jim Chetwynd at AFGL for recommending MODTRAN rather than LOWTRAN 7; P. Zvengrowski and K. Salkauskas, of the Mathematics Department of the University of Calgary, for helpful discussions with A.T.Y. of Hilbert space and functional analysis; Ted Ziajka of the Academic Computer Services of the University of Calgary, for providing A.T.Y. with an account on the IBM RS6000; Barbara McArthur, of the University of Texas, for providing help with *GaussFit;* members of the Working Group on Infrared Extinction and Standardization, particularly Ian McLean for commissioning the WG; and M.S. Bessell, T.A. Clark, George Rieke and Mike Skrutskie for useful discussions. Roger Heatley of Barr Associates provided much helpful information about interference filters. This project was supported through grants from the University of Calgary Research Grants Committee and the Natural Sciences and Research Council of Canada to EFM, for which we express gratitude.

References:

Bessell, M.S. and Brett, J.M., 1989, in: Milone, E.F. (ed.) *Infrared Extinction and Standardization,* (Lecture Notes in Physics, vol. 341). Springer, Berlin, p. 61.
Forbes, J.D., 1842, *Phil. Trans.*, **132**, 225.
Hodapp, K.-W., Rayner, J. and Irwin, E., 1992, *Pub. Astr. Soc. Pacific*, **104**, 441.
Jefferys, W.H., Fitzpatrick, M.J., McArthur, B.E. and McCartney, J.E., 1988, *GaussFit User's Manual,* University of Texas at Austin.
Johnson, H.L., 1965, *Astrophys. J.*, **141**, 923.
King, I., 1952, *Astron. J.*, **57**, 253.
Manduca, A., Bell, R.A., 1979, *Pub. Astr. Soc. Pacific*, **91**, 848.
Manfroid, J. and Heck A., 1984, *Astron. Astrophys.*, **132**, 110.
Milone, E.F. (ed.), 1989, *Infrared Extinction and Standardization* (Lecture Notes in Physics, vol. 341) Springer, Berlin.

Schwarzenberg-Czerny, A. 1991, *Astron. Astrophys.*, **252**, 425.
Volk K., Clark T.A.and Milone E.F., 1989, in: Milone, E.F. (ed.), *Infrared Extinction and Standardization* (Lecture Notes in Physics, vol. 341), Springer, Berlin, p. 15.
Wainscoat, R.J. and Cowie, L.L., 1992, *Astron. J.*, **103**, 332.
Young, A.T., 1974, in: Carleton, N. (ed.), *Methods of Experimental Physics,* vol. 12, *Astrophysics,* Part A: *Optical and Infrared* (Chapter 3), Academic Press, New York.
Young, A.T., 1988, in: Borucki, W.J. (ed.), *Second Workshop on Improvements to Photometry,* NASA CP-10015), NASA, p. 215.
Young, A.T., 1989, in: Milone, E.F. (ed.), *Infrared Extinction and Standardization,* (Lecture Notes in Physics, vol. 341), Springer, Berlin, p. 6.
Young, A.T., 1993, in: *Stellar Photometry, Proc. IAU Coll. No. 136*, eds. Butler, C.J. and Elliott, I., Cambridge University Press, p. 80.
Young, A.T., 1994, *Astron. Astrophys*, **288**, 683.
Young, A.T., Milone, E.F. and Stagg, C.R., 1993, in: *Stellar Photometry, Proc. IAU Coll. No. 136*, eds. Butler, C.J. and Elliott, I., *Cambridge University Press*, p. 235.
Young, A.T., Milone, E.F. and Stagg, C.R., 1994, *Astron. Astrophys. Suppl*, **105**, 259.

Liquid mirror telescopes

E.F. Borra, R. Content and L. Girard

Centre d'Optique, Photonique et Laser, Departement de Physique,
Université Laval, Quebec, Canada.

Abstract

Liquid mirrors allow one to build inexpensive zenith telescopes that can be used for robotic observations. I shall review the status of the liquid mirror project. A 1.5-m liquid mirror has been thoroughly tested and shows diffraction-limited performance. We have carried out observations with 1-m and 1.2-m LMT's that have led to published astronomical research. We have recently built a 2.5-m mirror. Interferograms show well-defined and straight interference fringes. The cost of a duplicate mirror would roughly be $40,000. I shall comment briefly on the field-of-view issue.

1. Introduction

It is straightforward to show (Borra 1982) that, in a rotating fluid, adding the vectors of the centrifugal and gravitational accelerations gives a surface that has the shape of a parabola. Using a reflecting liquid one therefore gets a reflecting parabola that could be used as the primary mirror of a telescope. We have carried out a feasibility study to determine whether liquid mirrors are practicable and summarize here the current state of the project. Interested readers should consult a recent paper (Borra et al. 1992) that discusses detailed optical shop tests of a 1.5-m diameter liquid mirror and describes the technology.

2. Liquid mirrors

The focal length of the mirror, L, is related to the acceleration of gravity, g, and the angular velocity of the turntable, ω, by

$$L = g/(2\omega^2) \tag{1}$$

For large mirrors of practical interest the periods of rotation are of the order of 10 seconds and the linear velocities at the edges of the mirrors range between 5 and 20 km/h.

The main interest in liquid mirrors resides in their low costs (one to two orders of magnitude less than a glass liquid mirror and its cell) and their very high optical qualities. Their main limitation comes from the fact that they cannot be tilted; therefore the accessible sky is limited by the field of view of the corrector and tracking must be carried out by means other than tilting the mirror. Present corrector designs allow correction over fields of view of the order of 1° giving access to a strip of sky 1 wide and nearly 360° long. This is actually quite a large surface of sky to study with a large mirror. Table 1 in Borra (1987) shows that 200,000 galaxies and 10,000 quasars are accessible to B=22 in a 1° wide strip of extragalactic sky ($b^{II} > 30°$). Networking several telescopes at different terrestrial latitudes would give access to most of the sky. Furthermore, the present corrector designs were developed when photographic emulsion was the standard detector, therefore they strive to give simultaneous correction over a nearly flat field. On the other hand, CCD detectors have small sizes, so that correction is only needed over a relatively small area. This allows correction to larger distances from the optical axis, as shown by Richardson and Morbey (1987). They carried out ray-tracing computations that showed that it is possible to obtain 0.2 arcsecond images at $\pm 7.5°$ off-axis. This is a large piece of sky. Considering that a conventional telescope is seldom used further than 45° from the zenith, a 15° field ($\pm 7.5°$) would give about 1/6th of the field accessible to a conventional telescope. We are presently beginning to systematically investigate the characteristics of correctors optimized for use with liquid mirrors, expecting to improve on the corrector described by Richardson and Morbey (1987), since they point out that their design is not optimized. In particular, we will strive to obtain a simpler design using warped metallic mirrors.

Modern technology now gives us alternate tracking techniques. For imagery, narrow-band filter spectroscopy, or slitless spectroscopy, one can use a CCD detector in the driftscan mode, store the information on disk and code the nightly observations with a computer. Imagery with a fixed telescope has been demonstrated by McGraw et al. (1986) and slitless spectroscopy by Schmidt et al. (1987). High- and medium-resolution spectroscopy can also be adapted to fixed telescopes. For example, Weedman et al. (1987) are planning to implement a fiber tracking system, that feeds the light to a fixed spectrograph, with a transit telescope. Indeed Borra (1987) has argued that essentially any type of astronomical instruments could be adapted for observations with a fixed telescope.

3. Optical shop tests

Fig. 1 shows the largest mirror that we have made. It has a diameter of 2.5 m and a focal length of 3 m. At f/1.2 this is probably the fastest large mirror ever made. Figure 2 shows an early interferogram taken with a scatterplate interferometer. Although the mirror adjustments were not optimized, the interferogram shows reasonably straight fringes indicative of a good quality surface. Since the Kilkenny workshop, interferometric tests of the f/1.2 2.5-m diameter liquid mirror carried out with a scatterplate interferometer (Borra et al. 1993) show Strehl ratios of order 0.6, close to the value of 0.8 usually taken to signify that diffraction limit has been reached. The mirror is certainly better than implied by the data because the interferograms were taken with 1/500-second exposures and the wavefronts therefore include the effects of seeing in the testing tower.

We have extensively tested a 1.5-m diameter f/2 liquid mirror with a scatterplate interferometer. The interferograms were observed with a CCD camera and analyzed with a computer program. Fig. 3 shows a typical wavefront, the statistics of the wavefront are given in the frame. The resolution on the mirror is 3 cm. The interferograms are observed with 1/30th of a second exposure time. We do not average frames because the mirror is liquid and the surface may change with time; as a consequence, our interferograms include seeing effects in the testing tower. Fortunately, our testing tower is in a basement room lined with thick concrete walls and has excellent seeing. We have analyzed several dozens and videotaped hours of interferogram observations to satisfy ourselves that the interferograms analyzed are indeed representative. The main conclusion of the interferometry is that a well-tuned mirror has RMS $\sim \lambda/20$ ($\lambda = 6228$Å) and a Strehl ratio near 0.8. This equals the performance of HST without spherical aberration.

Fig. 1: Photograph of the largest mirror built in our laboratory, having a diameter of 2.5 m and focal length of 3 meters.

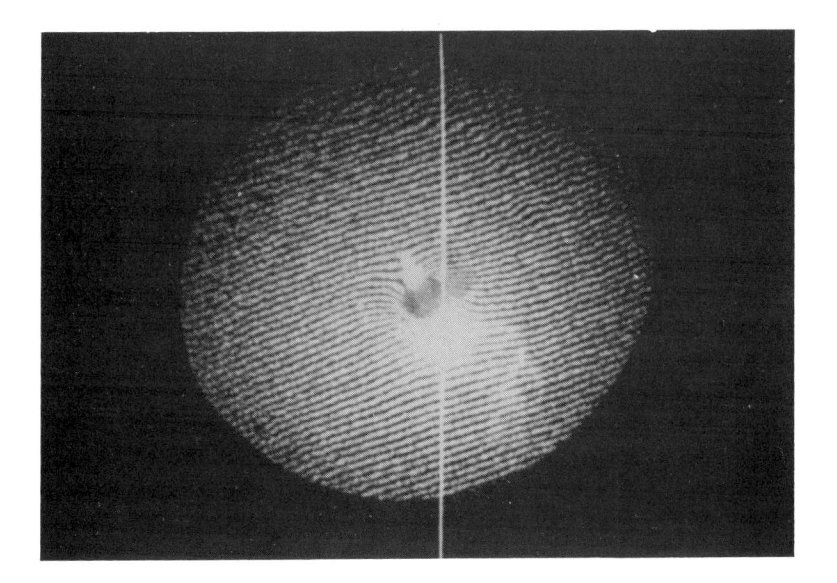

Fig. 2: Interferogram of the mirror shown in Fig. 1.

RMS = 0.046 λ
P-V = 0.444 λ

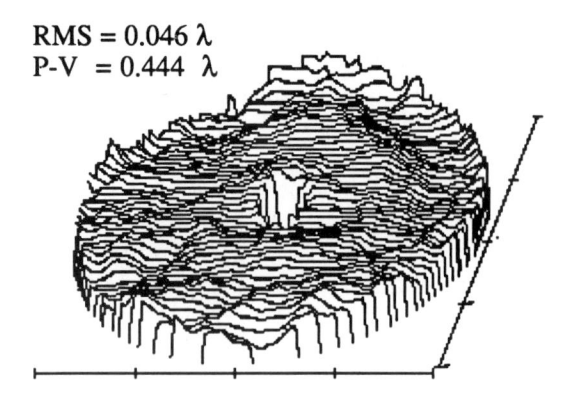

Fig. 3: Typical wavefront of the 1.5-m mirror that has been extensively studied. The statistics of the wavefront are given in the figure.

We have also measured the scattered light of the 1.5-m mirror. Scattered light measurements are very difficult and are rendered more so by our testing set up that has several auxiliary reflecting and refracting surfaces that introduce scattered light of their own as well as ghost images. Comparison of our Point Spread Function (PSF) to the PSFs of telescopes indicates that the scattered light of the 1.5-m and the 2.5-m are smaller.

4. Observing with liquid mirrors

At some stage, it is necessary to use these mirrors on the sky. A 1.2-m liquid mirror observatory was built and operated every clear night during the summer and fall 1986. During the fall of 1987, we operated a 1.2-m diameter LMT with an improved observatory building. The detector was a programmable 35-mm camera that registered star trails of 2-minute duration. The star trails give us information on the behavior of the liquid mirror during actual working conditions. We obtained 300 hours of data on film during 63 clear nights. The instrument performed well, especially considering that the frame and observatory were rudimentary. The evaluation of the data can be found in Borra et al.(1988). We also had a scientific project, looking for flashes in the sky and flares in stars, the results of which have appeared in Content et al. (1989). This paper is a milestone for it is the first time that astronomical research with a liquid mirror telescope has been published.

In collaboration with P. Hickson of the University of British Columbia, we have completed construction of a 2.7-m diameter liquid mirror astronomical observatory. It is a research instrument equipped with a driftscanning CCD. It will be used for a survey of quasars and galaxies. In collaboration with R. Sica of the University of Western Ontario, we have completed construction of a 2.7-m diameter liquid mirror LIDAR observatory. This is a research instrument equipped with a sophisticated and powerful laser that will be used for atmospheric research. These rank among the 15 largest telescopes in the world.

The 2.7-m LMT astronomical observatory has been operated for 3 months and the 2.7-m LMT LIDAR, used for atmospheric sciences has been in routine operation for 2 years.

Fig. 4 shows a CCD image taken with the astronomical telescope and 140 seconds exposure (the time it takes an object to cross the CCD). Notwithstanding the relatively bright sky, illuminated by the nearby city of Vancouver, the frame reaches R=21 (300 Å filter). With the testing of a 2.5-m LM and the first light of two 2.7-m LMT observatories we have demonstrated that large LMs are feasible.

Work on correctors optimized for LMTs has proceeded well. A practical corrector design consisting of a simple one-mirror active corrector capable of correcting within a 20-degree field has been made (Wang et al. 1994). Its field of view is small but sufficient for spectroscopy with optical fibers. A small prototype mirror has been made and successfully tested. Preliminary designs of a two-mirror corrector give PSFs useful for imagery within 10 arcminute patches inside 40-degree fields (Borra et al. 1994).

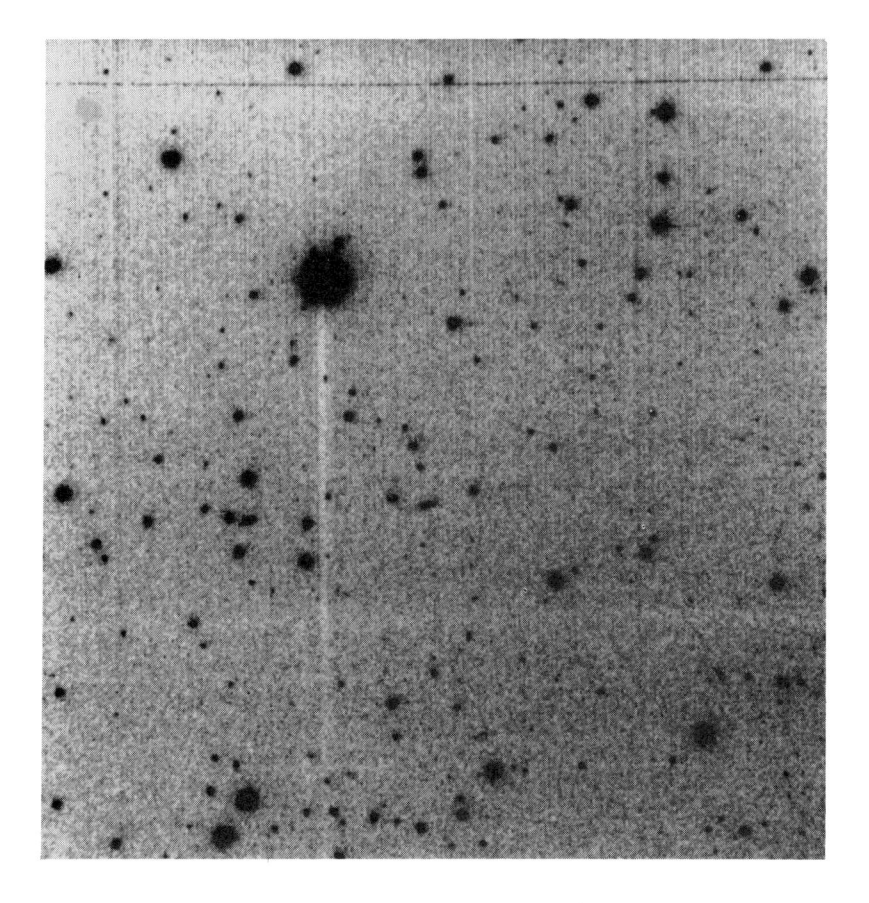

Fig. 4: Image obtained through a 300 Å filter centered at 7,000 Å with the 2.7-m UBC-Laval LMT. Numerous stars and galaxies in this 20 arcminute frame can be seen (Hickson et al. 1994). The faintest objects have R = 21.

5. Conclusions

Astronomical quality liquid mirrors are a fact. Indeed, at the measured Strehl ratios they are of "space quality". The 2.7-m UBC-Laval liquid mirror has begun to provide us with astronomical observations and will be a test-bench prior to building larger systems. Should someone want to build right away a liquid mirror observatory, we give the following advice. Begin by reading the Borra et al. (1992) paper that describes the tests and the technology. Content's 1992 Ph.D. thesis (in French) provides more extensive details. Design the observatory, telescope and instrumentation for a 2.5-m diameter mirror but begin by installing and operating a 1.5-m mirror, it will be extremely easy to build and will give the needed experience. After a few months of operation, replace the container of the 1.5m mirror with a 2.5-m mirror. We wish to point out that we certainly do not want to discourage anybody from building larger mirrors, but rather to give the best advice to build, quickly and with a minimum of trouble, a working research instrument. In conclusion, buy a commercial astronomical CCD and operate it in the drift-scan mode. Begin with a simple observational project involving imagery through wide-band or narrow-band filters. Finally, we will assist you in any way we can.

References:

Borra, E.F., 1982, *J.R.A.S. Canada*, **76**, 245.
Borra, E.F., 1987, *PASP*, **99**, 1229.
Borra, E.F., Content, R., Poirier, S. and Tremblay, L.M., 1988, *PASP*, **100**, 1015.
Borra, E.F., Content, R., Girard, L., Szapiel, S., Tremblay, L.M. and Boily, E., 1992, *Ap.J.*, **393**, 848.
Borra, E.F., Content, R. and Girard, L., 1993, *Ap.J.*, **418**, 943.
Borra, E.F., Moretto, G. and Wang, 1994, *submitted to A&A*.
Content, R., 1992, Ph.D. *thesis, Laval University*.
Content, R., Borra, E.F, Drinkwater, M.J., Poirier, S., Poisson, E., Beauchemin, M., Boily, E., Gauthier, A. and Tremblay, L.M., 1989, *AJ*, **97**, 917.
Hickson, P., Borra, E.F., Cabanal, R., Content, R., Gibson, B.K. and Walker, G.A.H., 1994, *Ap. J. Letters (in press)*.
McGraw, J.T., Cawson M.G.M. and Keane M. J., 1986, *Proc. SPIE*, **627**, 60.
Richardson, E.H. and Morbey, C.L., 1987, in: *Instrumentation for Ground-Based Optical Astronomy Present and Future*, ed. L.B., Robinson, New York: Springer-Verlag
Schmidt, M., Gunn, J., and Schneider, D., 1987, *Ap.J.*, **310**, 518.
Wang, M., Moretto, G., Borra, E.F. and Lemaître, G., 1994, *A&A*, **285**, 344.
Weedman, D., Ramsey, L., Ray, F. and Sneden, C., 1987, Bull. A.A.S., **18**, 956.

Parallelism in telescope and instrument control systems

I.G. van Breda

School of Cosmic Physics, Dunsink Observatory, Dublin 15, Ireland.

Abstract

By using techniques developed for parallel processing, it is possible to simplify control systems for telescopes and instruments. In particular, internal wiring inside instruments can be reduced, while less "glue" logic is needed for support microprocessors. Use of a distributed database simplifies the programs inside individual microprocessors, doing away with the need for an elaborate real-time operating system. System programming is also simplified, since much of it takes the form of setting up of configuration and database files and so requires less specialist programming effort.

1. Introduction

When mention is made of parallel processing, it usually conjures up an image of a large powerful computer, carrying out vast calculations on computationally intensive tasks such as three-dimensional problems in fluid dynamics. Yet parallel processing is very commonplace, even in ordinary desktop computers: apart from the central processing unit (CPU), there will, for example, be processors to help in generating the screen display, transferring data to and from disc, sending data to a printer and generating sounds. These all assist the CPU by off-loading tasks, so allowing it to concentrate on the main job of running a program.

In telescope, instrument and other control systems, the parallelism is more obvious, since individual instruments may have their own control microprocessors and, indeed, will often have several. For an actively controlled 8-metre mirror, the number could be in the hundreds.

2. Types of parallelism

There are many ways in which parallelism can be implemented in electronic circuitry, depending on the nature of the data and the types of operation that are required.

2.1 PIPELINING

Pipelining is commonly employed in most of the more advanced current generation of microprocessors. As each instruction is extracted from memory, it is passed into a pipeline, where it is successively decoded and executed. Any results are then passed back to memory. As each instruction is passed on to the next stage, a new one is inserted into the pipeline so that several instructions are being processed at different stages at any one time. Pipelining works best when the instructions are in a long stream without jumps, in order to avoid breaking up the flow of the pipeline.

2.2 ARRAY PROCESSING

Another way in which parallel operation may be achieved is in array processing, where many identical operations are carried out on the different array elements. This is ideal for image processing calculations. For example, in massively parallel processors, there may be many single-bit processors connected in parallel and, because they cover many pixels simultaneously, even very complex floating-point operations can be carried out highly efficiently over whole images. Pipelined floating-point processors may also be used to speed up image processing calculations.

2.3 DATA-DRIVEN SYSTEMS

A third form of parallel architecture is described as data-driven. In this arrangement, it is the availability of data that drives program execution; the approach is particularly well suited to control system applications.

In the classical so-called von Neumann computer architecture, a processor carries out a sequence of instructions, one at a time, jumping around the program as it goes. In a data-driven system, a program may be divided up into a number of independent "von Neumann segments", each of which must be executed in sequence internally, but which may be initiated at any point in time, so long as the necessary data is ready. For example, the segment of pseudo code,

$$\text{If } x > 0$$
$$y := \sin(x)$$

$$\text{Otherwise}$$
$$y := -\sin(x)$$

requires an input data value x and returns an output value y.

Immediately the value of x is known, the value of y may be calculated for use by some other segment in the program. The program completes when all those segments where input data has changed have been processed. Of course, this procedure needs to be convergent, as a calculation could be done more than once, if it has more than one varying input, and precautions must be taken to avoid problems from this source.

A less simplistic example is shown in Fig. 1. Here data are used to assemble a FITS tape and disc file. Complete header data is available at the end of an exposure, including the name and co-ordinates of the object, time and length of observation, image size and various other similar parameters for the observation. These can be assembled into the header before the detector is read out, but the complete file can only be written to disc and tape when the image data have become available.

An important point to note about this is that time is necessarily used up in the execution of each segment, even though many segments may be executed in parallel. This use of time is also an essential component of all causal logic: we say, "if so-and-so is true, *then* such-and-such happens". Equally, time must be allotted in order to move a motor, because of its inertia, or to allow exposure time for an image from the telescope.

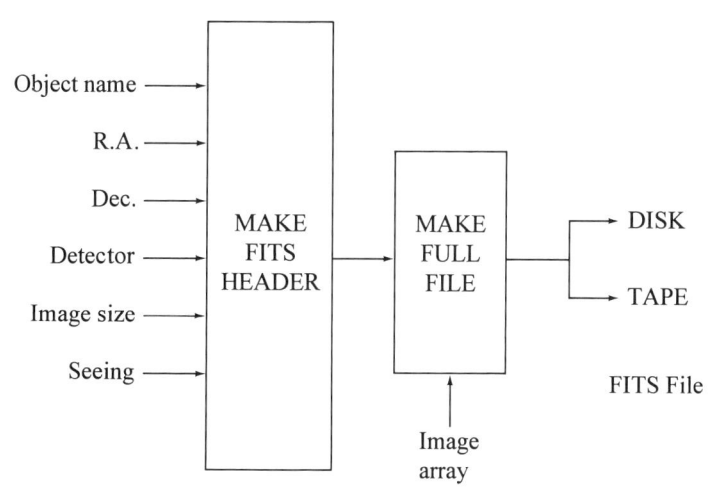

Fig. 1: Building a FITS file from header parameters and image data.

A particularly interesting instance of a data-driven parallel processor is the processing part of the brain (see, for example, the detailed account in Alberts et al. 1989), which may need to do many things in parallel, for example, when driving a car or playing a musical instrument.

The parallel structures of the brain are well adapted to the survival situation, enabling us to recognise, say, a face much more quickly than a Cray computer can, even though the response speeds of individual neurons are quite slow: responses vary according to function, but are typically a few milliseconds. When a scene changes, inputs from the visual sensory cells are fed to neurons in the visual cortex, where they fire many other neurons. These are connected in complex ways, allowing the scene to be interpreted stereoscopically very quickly. By contrast, the brain is not well-adapted to floating-point arithmetic, since that has little to do with survival.

3. Electronic implementations

As indicated above, there are several general ways of introducing parallel operations into electronic hardware and many variants within each method. We shall discuss just a few of them, especially as they affect control systems.

3.1 NEURAL NETS

Neural nets are intended to mimic the brain by electronic means. However, they fall far short of the complexity occurring in biological organisms. For example, neural connections in the brain involve complex and subtle analogue signals, with majority logic connections that can be both excitatory and inhibitory; a single neuron, of which there are around 10^{11} in the human brain, may itself have many thousands of connections. Also, the brain is three-dimensional by nature, whereas present day electronics are essentially two-dimensional, but with three-dimensional wiring and printed circuit connections. In practice, since the electronic problems are so great, neural nets are usually simulated in a digital computer.

One motive for developing this type of computing is to handle those cases where a process is not well understood but the required results are known, as in automatic classification of galaxies. Another application would be in image photon counting to separate overlapped events; in principle, a neural net can be trained to do this using simulated images covering a variety of overlapping event situations, with each event having known co-ordinates.

However, where algorithms are well understood, it is always better to use direct methods than to use neural nets. For example, it is better to write a program to control telescope tracking using spherical astronomy than to tell a neural net, "follow that star".

3.2 TRANSPUTERS

One of the first attempts to produce a microprocessor specifically aimed at parallel processing was the Inmos (Thomson) Transputer. Transputers are Reduced Instruction Set (RISC) processors, usually with four pairs of bidirectional links, which can be connected together into a variety of processor array configurations. A special language, Occam, has been developed for programming parallel systems, although other more conventional languages have also been extended for the purpose.

Although the Transputer seeks in some sense to copy the neuron, its action is strictly digital and the connections are very limited by comparison. While serial links simplify interconnections, they do impose certain bandwidth limitations when transferring data from one processor to another, even though data rates are very fast at a 5—20-MHz bit rate.

Transputers have a simple internal structure and the smaller versions are very economical. This makes them ideal for control systems, where individual processors can, for example, be allocated to individual motors in an instrument without significantly increasing the budget required. In control applications, Transputers have the great advantage over many other microprocessors in needing very little "glue" logic to operate, so helping to keep costs down, while interconnections can all be serial, instead of the more usual parallel backplane for connecting interface cards.

3.3 NEURON CHIP

More recently a chip, called the Neuron, has been developed specifically for distributed control systems in buildings (see the LonWorks Product Line Brief from Echelon Corp.; chip manufacturers are Motorola and Hitachi). This also requires very little glue logic and is designed to be cheap enough to build into individual lights and switches (unfortunately the development systems are currently rather expensive).

Instead of using individual serial connections, the Neuron chip is connected to a common bidirectional twisted-pair serial link at bit rates of 78 Kbps or 1.25 Mbps. Like Ethernet, it relies on detection of message collisions by the processors to avoid loss of information. This link is rather slower than those of the Transputer, but it is sufficiently fast for control of electromechanical devices. Radio and electrical mains-borne signals can also be used at reduced data rates.

4. Control system structures

Telescope and instrument computer control systems can be considered to have gone through several generations of development, starting initially with parallel wiring connections to a central computer. Later, microprocessors were introduced for control of individual instruments, with connections made using serial links and high-level messages (van Breda and Parker, 1983). Later still, and to simplify the connections further, networking was introduced, as on the William Herschel Telescope, which uses Ethernet to connect the instruments together (Parker et al. 1983).

However, there are disadvantages to the conventional networking approach: message protocols can become very complex and Ethernet needs an overhead on every message of around eighty bytes. Since wiring connections to Ethernet are expensive, instruments will usually be interfaced using RS-232 serial links, for reasons of economy, and the system response can then become very slow. A further difficulty is that the software needed to manage the links on the microprocessors also becomes rather complex.

More recently, use of parallel systems has been suggested for telescopes on the grounds that both hardware and software can be further simplified, while at the same time greatly improving response speeds (van Breda et al. 1990).

A suitable structure for a telescope control system using Transputers is shown in Fig. 2. A single fibre-optic link is run between Transputers at the major nodes in the control system, including a connection to the main system computer; a spectrograph might be considered complex enough to have its own major node. Since the link is bidirectional, closure of the loop is optional, but provides a measure of redundancy, should things go wrong.

Each major node Transputer siphons off messages for instruments attached to it, while allowing messages to other nodes to pass. Since messages must always pass through the Transputers, this topology helps to keep down the number of processors in the chain for any particular message. However, there are few problems with speed, since the links operate at rates that are three orders of magnitude faster than the usual 9600 baud connections to Ethernet and also messages can be a further order of magnitude shorter, due to simpler protocols.

Neuron chips could also be used in place of Transputers within each individual instrument, if desired.

For reasons of speed, detector data are best fed directly through a separate link to the interface Transputer attached to the system computer bus, although a single link will cope readily with both commands and data from a single CCD. Where speed is particularly important, Transputers within instruments and detectors may be connected directly together to reduce the number of connections through which messages must pass.

Interconnection between electronic sub-components is greatly improved since it is no longer necessary to use a parallel backplane with relatively complex interface cards. Instead, serial links with simpler wiring and electronics make it much easier to take an instrument apart; sub-components can then be tested more readily at the workbench for debugging purposes, where they can be re-attached to the parallel network, if needed.

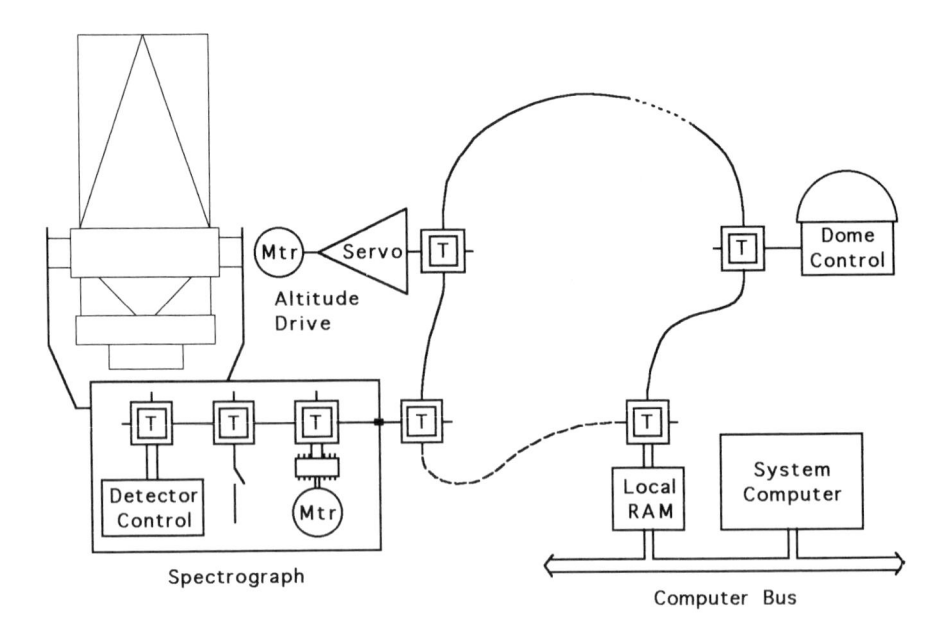

Fig. 2: Schematic parallel telescope and instrument control system. Long-distance communication is by fibre-optic link, short-distance by high-speed electrical serial link. Boxed Ts denote Transputers.

5. Distributed databases

One of the problems encountered with a conventional network is that it is relatively complex to program. Special programs must be written for each instrument using a multi-tasking real-time operating system, which can involve quite complex status checking and protocols. Thus extending the system by adding a new instrument becomes quite a difficult task for a specialist programmer.

Software simplicity is an important goal for all telescopes, but especially for robotic telescopes which must operate unattended. Operating systems have grown considerably in complexity recently, with little sanction on their size: the computer does not grow bigger or use more electrical power when the operating system grows, leaving little incentive to keep things simple. Nevertheless, simplicity is important as it improves reliability.

The parallel approach can greatly simplify the programming task by using the concept of a distributed database (van Breda et al. 1990; Echelon Corp. literature), a concept that has been growing in popularity as a means of controlling parallel computer systems for other purposes.

In this scheme, we treat the state of the system as a data-driven database. The state at any one time is represented by a number of data objects whose values are distributed throughout the system. Each data object may be supplied by just one source, with copies of that object being passed to those nodes that need to know about it; often this is just one node.

Two general schemes may be used, one where the database is totally distributed and one where there is a "central nervous system". Both are very closely equivalent. The second scheme is attractive for telescope/instrument control, since a copy of the complete database is necessary for observing, checking correct progress or debugging; it is also convenient to have a central machine for initialisation. Except for large images, a complete copy of the database can be held in two-port RAM located between the system computer and the main interface Transputer (Fig. 2).

Items that might be considered as data objects are:

(i) filter, aperture and grating configurations;
(ii) frequency and phase for a nodding secondary;
(iii) event flags, e.g., to start an exposure;
(iv) object co-ordinates;
(v) temperature and humidity sensing;
(vi) motor ramping parameters;
(vii) driver programs;
(viii) detector output (photon counts and images);

along with many others.

The system operates by passing data objects between nodes, typically when data changes at the source. For example, a change occurring in a temperature sensor should cause transmission of the corresponding data object. Each message on the network must contain a "route map" identifier that determines which processors the message must traverse to reach its destination(s). Such an arrangement makes reconfiguration relatively easy, an important factor for remote telescopes when a failure is not catastrophic.

An ideal way to implement the above scheme is for each received data object to initiate an associated routine in the receiving node. For example, a newly requested position for a motor should initiate the routine to drive the motor to its new position and, in turn, issue its own encoder reading object to send back to the system computer. When the new position is received by the main interface Transputer, it triggers a routine that passes the value to the instrument configuration table held in the two-port local RAM.

Since the control processors are dedicated, the central system computer does not need to have a real-time capability and even a Unix workstation will do the job. Neither is a separate telescope computer needed: Transputers in the drive servos can be fed with tracking splines by the system computer, which can easily keep up with the required calculation rates.

Only a limited set of programs is needed for such things as motor drivers and sensor readouts. Generally, these programs can be quite simple, although certain specialist applications, such as photon counting and CCD readout, will need to be more sophisticated (Smith 1989, Waltham et al. 1990, Carter et al. 1990).

Setting up the system then involves: naming each Transputer and specifying which programs are to run on which processor; specifying the way in which Transputers are connected together, i.e. which link on one Transputer is connected to which link on another; specifying the data objects in the database, including the source and destination(s) of each one. Later, when changes are needed, they can often be made by non-specialist programmers, since configuration changes are usually all that will be required. Nevertheless, the basic routines inside individual processors, which are less prone to change, will still need specialist attention and must, naturally, be thoroughly debugged.

References:

Alberts, B., Bray, D., Lewis, J., Raff, M., Roberts, K. and Watson, J.D., 1989, *Molecular Biology of the Cell*, Garland, 2nd ed., Chapter 19.
Carter, M.K., Cutler, R., Patchett, B.E., Read, P.D., Waltham, N.R. and van Breda, I.G., 1990, in: *Proc. SPIE Conf., Instrumentation in Astronomy VII*, **1235**, 644.
Parker, N.M., van Breda, I.G. and Martin, R., 1983, in: *Proc. SPIE Conf., Instrumentation in Astronomy VII*, **445**, 505.
Smith, R.M., 1989, *Microprocessors and Microsystems*, **13**, 149.
van Breda, I.G. and Parker, N.M., 1983, *Microprocessors and Microsystems*, **7**, 203.
van Breda, I.G., Newton, G.M., Johnson, A.N. and Waltham, N.R., 1990, in: *Proc. SPIE Conf., Instrumentation in Astronomy VII*, **1235**, 438.
Waltham, N.R., van Breda, I.G. and Newton, G.M., 1990, in: *Proc. SPIE Conf., Instrumentation in Astronomy VII*, **1235**, 328.

Editor's summary

M.F. Bode

Astrophysics Group, Liverpool John Moores University
Byrom Street, Liverpool L3 3AF, UK.

1. The present state of the art

Automated and robotic telescopes currently take a variety of forms in projects with varying levels of sophistication and ambition. These range from conversions of existing telescopes (e.g., the 60-cm Belogradchik and 14-inch Kotipu Place telescopes), to those developed as robotic telescopes from the outset (e.g., APTs, the Bradford Robotic Telescope, the 3T1M Project, the Argentinian Robotic Telescope etc.). In both cases we are now moving on from telescopes using photomultiplier tubes to those in which the CCD is the primary detector, and from simple target acquisition (for example using square spiral search techniques) to pattern recognition and autoguiding, again via CCD cameras.

The advantages of CCD detectors over multiplier tubes are obvious. As outlined by Alexei Filippenko in relation to the Berkeley Automatic Imaging Telescope (BAIT—Filippenko 1992) these can be summarised as follows: (i) They have relatively high quantum efficiency with respect to photomultiplier tubes enabling fainter objects to be observed in a given time, (ii) the effective entrance aperture is defined by the image itself and can therefore be matched more flexibly to seeing conditions, thus decreasing the sky contribution optimally in a given exposure, (iii) similarly, a local background (e.g., in the case of a supernova lying against its parent galaxy) can be fitted and subtracted, (iv) in crowded fields stellar profile fits can be used, (v) imaging allows differential photometry to be utilised very effectively in marginal conditions thus extending the number of nights per year on which photometry can be performed, (vi) moving objects such as comets and asteroids can be monitored, and finally (vii) in addition the option remains to put some form of dispersive element in the optical path and therefore use the CCD as the detector for a spectroscopic mode of operation. Coupled with autoguiding, the use of CCD detector technology makes photometry to $V=21$ feasible for 1-m class telescopes in long exposures. Thus the range of projects in which such telescopes can partake is vast.

2. Current challenges

The use of CCDs for target acquisition, autoguiding, and observation of course presents its own problems. Primary among these is the deluge of data that ensues. Nowadays each CCD frame can potentially generate several megabytes of data. On a typical night a robotic telescope may take several hundred such frames. Although it may well be that only sub-areas of each of these will be stored and analysed, this still means that data storage can best be done on optical discs. Ideally the data should also be fully reduced on-site to minimise the data transmission rate required back to the telescope's remote home base. Such problems are well appreciated by those currently building CCD-based robotic telescopes (see, for example, John Baruch's contribution in this volume).

If such telescopes really are to work in crowded fields when performing photometry on very faint objects, then reliable pattern recognition is essential. No robotic system can afford to have the credibility of its data questioned. The Bradford project is an example where particular attention is paid to this problem. Similarly, David Kilkenny's contribution emphasised the way in which solutions to reliable acquisition with a CCD imager have evolved as the SAAO telescope project has progressed.

In his paper, Ian van Breda concentrated on the application of parallel processing to simplify control systems for telescopes and their instruments. The whole of this area is obviously of primary importance to the efficient and effective running of a robotic telescope. The analogy that van Breda draws with the human brain performing many tasks in parallel, for example when driving a car or playing a musical instrument (or even controlling a large telescope!) is an excellent one.

As the era of networks of robotic telescopes approaches, attention is turning to the fundamental bases of photometry. Chris Sterken has again drawn a wonderful analogy between the development of the modern metric system and the way in which photometric systems have and should evolve to meet this challenge. Tarmo Oja has even called our attention to the fact that so-called photometric standards may not be so standard after all. As robotic telescopes enable us to work to millimagnitude precision it is obvious that a close look at standardisation of photometric systems needs to be undertaken as a matter of some urgency. It is recognised, however, that there is great inertia in the astronomical community that will need to be overcome before real progress can be made.

3. The near future

From a basic engineering point of view, we have probably witnessed the evolution of the first two generations of robotic telescopes. Those employing photomultiplier tubes as their primary detectors were first generation and now, as noted above, several projects are underway which rely on CCDs and are therefore much more technically demanding. The third generation of robotic telescopes will push technological frontiers even further. The Future Small Telescope concept for instance, involving a collaboration between the Royal Greenwich Observatory and Liverpool John Moores University, exemplifies what David Crawford would call a new generation small telescope. At its heart is technology transferred directly from large telescope projects such as the 4.2-m William Herschel and subsequently upgraded and improved to provide the highest engineering and optical quality on 1-m to 3-m class telescopes which will then have the option of being robotically controlled.

As an example of radical new ideas which may be employed we have seen the progress of Ermanno Borra's group in producing very large and efficient high-quality liquid mirror telescopes, at least three of which are already producing scientific results. As Borra emphasises, although by their very nature such instruments cannot access as much of the sky as a conventional telescope, the strips of sky they can observe still contain vast numbers of astrophysically interesting objects, and such telescopes must have great potential for surveys down to very faint limiting magnitude.

The way forward with individual telescopes may lie in the type of observations they are able to perform. For example, we have already mentioned the possibilities of spectroscopy, perhaps involving fibre-feed systems—although this obviously adds complexity to data reduction on-site. Robotic telescopes in the next decade will also undoubtedly be used for observations at wavelengths outside the visible, particularly in the infrared.

However, as emphasised by Milone et al. in their contribution, particular attention needs to be paid to some of the fundamentals of infrared photometry. In addition, problems of cryogen supply and mirror cleaning will need to be solved before further progress can be made.

Without doubt one of the most important developments will involve networking. This will allow many exciting problems in astrophysics to be addressed more thoroughly than has previously been possible. However, as alluded to above, the efficient and effective operation of such networks will present their own technical challenges in terms of the homogeneity of the photometric systems used. One of the most exciting prospects is Querci et al.'s 3T1M project involving the placing of telescopes on very remote sites, but using simple and robust technology. This project is similar in some respects to placing such observatories on the moon. The use of three telescopes at each site would undoubtedly produce some of the highest-quality photometric data yet acquired. Underpinning such networks of telescopes will be the information "super highway" we hear so much about.

4. Rewards of the venture

Over the next few years we can look forward to robotic telescopes enabling us to make great strides in several areas of observational astronomy. The range of possible programmes is wide. What follows is a sample of some of the areas ripe for exploitation, drawn from a variety of sources including a survey of interest in the UK astronomical community for the establishment of an "Automatic Monitoring Telescope". This survey resulted in the identification of around 80 separate projects requiring the unique capabilities of a robotic telescope.

4.1 SOLAR SYSTEM

In the Solar System robotic observatories would be ideally suited to observations of **asteroids** to determine rotation rates etc. Photometric variability can also be related to the changing phase angle of reflected sunlight enabling the surface physical structure to be explored. As pointed out by Filippenko (1992) previous observations have tended to be biased towards bright (large inner belt) asteroids. Robotic telescopes offer the prospect of sampling the smaller and/or more distant population of such objects. Similarly, there has been a natural bias towards observations of **comets** when they are close to the sun and/or unusually active. As he points out, there is in fact little data on inactive comets, or active ones greater than a few AU from the Sun, and our perception of comets is therefore likely to be biased as well. Following a subset of comets over many years and trying to observe in detail their normal behaviour is a task ideally suited to the robotic observatory.

There are still a number of outstanding problems associated with the orbits of the **inner satellites of the outer planets**. A significant number are within a few arc minutes of their parent bodies and so are well-suited to CCD studies. With orbital periods ranging from a few hours to around a day, an observational strategy of taking a few observations per night for a period of a week, repeated after a few months is called for.

4.2 STELLAR ASTRONOMY

The principal mechanisms for variability in **pre-main-sequence stars** appear to be flare activity, changes in photospheric temperature (possibly caused by starspots) and variable extinction caused by changes in circumstellar dust opacity, due either to condensation or to clumpiness in an orbiting edge-on circumstellar disc. In order to provide an effective diagnostic for these mechanisms in individual stars it is necessary to investigate both light curves in individual passbands and colour changes over the optical spectrum from U to I. Simultaneous Strömgren and (particularly) Hβ photometry is valuable, notably in cases where spectral variability occurs. Work done from conventional ground-based telescopes generally contains epochs in which the stars are observed intensively separated by long periods without any data. To compare observational data with models based on the above mechanisms a single measurement per week over a period of months is required in some cases, and this is clearly very difficult with current methods of scheduling time on conventional large telescopes.

We have already seen in the contribution of Hall and Henry that the well-sampled and extensive data sets generated by robotic telescopes are ideal for monitoring **magnetically active variables**. Activity includes flare events, starspot cycles and related chromospheric/coronal phenomena resulting from enhanced localised magnetic fields on stars which are younger, and more rapidly rotating than the Sun, or in such stars forced to rotate quickly because of tidal effects in close binaries.

Theoretical models of **pulsating stars** can now predict many of the phenomena observed in their light curves, and hence determination of stellar masses can be made for direct comparison with evolutionary tracks. Monitoring of pulsating stars, which are often multi-periodic in the range of tens to hundreds of days, is largely left to amateurs at present who obviously can rarely provide the accuracy of photometry, in multiple colours, that a professional instrument can deliver and that theorists require. The lack of appropriate data is particularly apparent in the study of Helium stars, RCBs, RV Tauris, a variety of subdwarfs, and Type II Cepheids. In the case of RCB stars, as well as undergoing pulsations, these objects are subject to unpredictable, sudden drops in brightness following the ejection of grain-forming material, and the rapid formation of carbon dust. Again these deep minima are generally monitored by the amateur community, but a detailed understanding of the grain formation process—one of the few astrophysical situations in which grain formation can be observed in real time—requires frequent regular monitoring, coupled with the ability to react quickly to sudden light curve changes. The potential for the application of robotic telescopes to detailed studies of these objects, and the related UU Her stars, is clearly demonstrated in Don Fernie's studies of the brighter members of the class as reported earlier in these pages.

Dwarf novae are interacting close binary systems with binary periods of less than two days, and mostly in the range 2 to 10 hours. They provide the nearest examples of accretion discs available to us to study, and this has immediate application to many areas of astrophysics. As with supernovae and classical novae discussed below, the outbursts of dwarf novae are unpredictable. Again, much of the observational data on visual light curves comes from amateur observations and long-term multi-colour high photometric accuracy monitoring of dwarf novae is therefore an ideal project for a robotic telescope. A detailed and systematic monitoring of these objects, which have recurrence time of outbursts typically of 1 to 6 months, will certainly lead to a dramatic increase in our knowledge of the behaviour of accretion discs, and the mass transfer process in general.

Classical novae are also short-period interacting binary systems, having high luminosity outbursts of an apparently unpredictable nature every 10^4 to 10^5 years. On average around three classical novae are discovered per year brighter than magnitude $V = 10$, and over 200 of these systems are now known in our Galaxy. Surprisingly little observational data are available immediately around maximum, and the subsequent following of the decline is generally erratic. Most observations of the post-outburst light curve are heterogeneous combinations of data taken at various observatories with a range of instruments. This generally precludes attempts to study in detail the occurrence of small amplitude light curve oscillations which may be associated with fluctuations in mass loss rate from the surface of the white dwarf star in the binary. Scheduling simultaneous multifrequency observations has also proved extremely difficult despite the fact that such studies, particularly during the dust formation phase, have proved especially fruitful. Robotic telescopes are ideal for the rapid response, long-term follow-up, and co-ordinated observations that are necessary for us to understand fully the later phases of classical nova outbursts. It would also serve us well to monitor classical novae at quiescence in projects aimed at the determination of orbital periods and monitoring dwarf nova-like outbursts. These objects have also been observed in our near-neighbour galaxy M31 since the early days of this century. Recent CCD imaging from conventional telescopes has led to a revision of the rate of classical nova outbursts in the nuclear regions of M31, and hence to the overall accepted rate for the galaxy. This in turn has ramifications for the inter-outburst timescale of these objects, and their total contribution to the enrichment of the interstellar medium with heavy elements and dust grains. Finally, novae are an important distance scale indicator, via the maximum magnitude/rate of decline relationship. To test fully the idea that these objects are in fact almost the ideal distance indicator in the local group of galaxies, projects aimed at securing an enhanced set of data on extragalactic novae are required. A robotic telescope with imaging capability would greatly extend and enhance work in these areas.

A class of object related to classical novae is the **symbiotic stars**, of which around 200 examples are known. Despite their importance in several branches of astrophysics (for example the evolution of binaries, mass loss by red giants, accretion processes) the nature of their hot components, and the cause of outburst are still controversial in many cases. An extremely important parameter needed in nearly all physical models is the orbital period, but this is only known with any certainty in a small percentage of the total number of objects. In some cases this has been derived via photometric periods. However, this relies largely on data supplied by the amateur community. A robotic monitoring telescope would be the best hope for making progress in determining periods (which are typically of the order of months to years) photometrically for any significant number of objects, as well as for confirming those periodicities which are already suspected.

The luminous **stellar X-ray sources** have been known for around thirty years, and they have greatly advanced our understanding of neutron stars and provided the most compelling evidence for the existence of stellar mass black holes. Such a compact object can produce copious X-rays whether its companion is an early-type star (OB supergiant or Be star) or a low mass star (analogous to the dwarf novae we have discussed above). Robotic telescopes would be particularly valuable in studying the long-term behaviour of the high mass systems, and also some of the low mass binaries. The early-type systems can have long periods and eccentric orbits, with accretion on to the neutron star occurring only near periastron. It is essential to monitor these objects over large fractions of the binary cycle in order to follow the interaction of the compact object and its more massive companion. The Be star systems, for example, all have periods longer than about 2 weeks, most lying between 30 and 150 days with one having a period as long as 550 days.

There are many other types of variable stellar source that will lend themselves to detailed investigation by robotic telescopes. These include the **Hubble—Sandage variables** which are the most luminous stars known and **Wolf—Rayet stars** in which are found examples of interacting binaries in eccentric orbits where interacting winds can produce effects from variable non-thermal emission to copious dust formation. Some of the most ambitious stellar projects are related to the development of networks and observations with millimagnitude precision. These include **astroseismology**, as described in the contributions of Querci et al. and Garcia et al. in this volume, and the **search for planets** around nearby stars (again see Querci et al.).

4.3 EXTRAGALACTIC PROGRAMMES

The ability of CCD-based photometry on such telescopes to reach 21st magnitude means that **supernovae** in the Virgo Cluster can be followed for around two years from maximum. As pointed out by Filippenko (1992) the acquisition of precise and long-term light curves for these objects will aid in our understanding in the physics of the outburst, and also enable us to use supernovae more reliably as standard candles in defining the distance scale of the Universe. Already the Berkeley Automated Supernova search, using a 20-inch robotic telescope at the Leuschner Observatory, has led to the discovery of a few tens of extragalactic supernovae.

As emphasised in the paper by Ian van Breda, application of robotic observatories to observations of **active galactic nuclei** is likely to prove very important in unravelling the nature of their central engines. Here high-quality precision photometry is required to monitor both short-term and long-term variations. In other words, observations will need to be made on time-scales from a few times a night to months and years in order to begin to understand the physical nature of these objects more fully. Similarly, for investigations of **gravitational lens** sources long homogeneous data sets with frequent sampling are required. Such data will then give accurate determinations of the phase lags between different components of the lensed image. The prize here is the definitive determination of the Hubble constant, H_0.

4.4 WIDER PROGRAMMES

Co-ordinated observations with the new generation of space observatories (for example ROSAT, the Infrared Space Observatory, the Hubble Space Telescope, the Advanced X-ray Astronomical Facility etc.) will undoubtedly flourish in the era of robotic observatories. The further development of the SCOPES initiative, as described in these pages by John Butler and his co-workers, will aid in both the co-ordination of multifrequency campaigns and also in the realisation of the full potential of network ground-based telescopes that we shall certainly see evolve in the coming years.

Robotic telescopes are ideal tools for enhancing **science education** and the **public understanding of science** in general. Already the Berkeley group are well on the way to using their robotic telescopes for undergraduate programmes in astronomy, thereby overcoming the great problems posed by the cloudy nights at observatories to which students are inevitably exposed and which inevitably lead to frustration. A robotic observatory gives such students, and potentially schools and colleges, access to real scientific data, and if necessary these telescopes can in fact be run remotely in real time. A robotic telescope operating in the United States for example could be run in this way by students in Europe during their daylight classes.

Robotic telescopes will be used more and more to perform sky survey **follow-up work**. For example, the recent UK Schmidt Telescope/COSMOS survey which aimed to discover blue stellar objects in the southern hemisphere required follow-up photometry and spectroscopy to relatively faint magnitudes to identify the type of blue object selected (an analogy may be drawn here with the ROSAT EUV survey). Such objects range from hot subdwarfs through blue horizontal branch stars to white dwarfs, cataclysmic variables, and bright quasars. Of special interest are the pulsating degenerate stars. The discovery of more of these is important for two reasons: (1) the pulsation provides more information about the structure of the star; and (2) changes in period enable checks to be made on the white dwarf cooling timescale. Accurate photometry with a robotic telescope would greatly aid in the discovery and subsequent detailed investigation of these objects. The Whole Earth Telescope project, a forerunner of GNAT, has shown how valuable a network of telescopes can be in this regard.

Programmes involving **surveys** and routine but **large-scale fundamental photometry** will undoubtedly become the domain of robotic telescopes. We have already been given evidence of the preliminary work relating to calibration of Sky Survey fields by the St Andrews group in the contribution of Hill et al. in these pages. In addition, such telescopes will prove invaluable for **patrol work** in general, including searches for supernovae in which, as noted above, the Berkeley group have already proved the utility of such facilities, and searches for earth-crossing asteroids as proposed in the "Spaceguard" project.

5. Closing remarks

The automation of telescopes has come a long way since the days in which the 72-inch telescope at Birr Castle was operated remotely via verbal instructions from the Earl of Rosse to his servants to move the telescope in its restricted fashion. We are in fact on the brink of a revolution in the way astronomy is conducted from ground-based observatories. In the medium-term, such telescopes will complement our suite of conventional large instruments. In the long-term they may completely supersede them. Virginia Trimble was, however, correct to draw attention to the pitfalls of over-reliance on a "black-box" approach whereby some of the serendipitous discoveries of recent times (for example pulsars) might not have been made (Trimble 1992). Nevertheless, the overwhelming advantages of such telescopes are clear and we are rapidly moving from an era in which the robotic telescope community was very insular to one where results from these telescopes are presented at mainstream astronomical meetings. Professional astronomers are increasingly directly involved in projects not only to use the data but also to construct the telescopes themselves. I strongly suspect, however, that a similar workshop in five years' time will concentrate more on the wealth of significant scientific discoveries that will undoubtedly flow and less on the technical development stage through which we are now passing. The future of astronomy with robotic telescopes is exciting indeed.

References:

Filippenko, A.V., 1992, in: *Robotic Telescopes in the 1990s*, ed. A.V. Filippenko, ASP Conf. Series, Vol. 34, p. 55.
Trimble, V., 1992, *ibid*, p. 359.

Appendix: List of Participants in the Kilkenny Workshop

Mr E. Ansbro	Fastnet Observatory, Co. Cork, Ireland
Dr A.P. Antov	Belogradchik Astronomical Observatory, Sofia, Bulgaria
Dr J. Baruch	University of Bradford, Bradford, U.K.
Prof M.F. Bode	Liverpool John Moores University, Liverpool, U.K.
Prof E.F. Borra	Université Laval, Quebec, Canada
Dr E. Budding	Carter Observatory, Wellington, New Zealand
Dr C.J. Butler	Armagh Observatory, Armagh, N.Ireland
Dr D.L. Crawford	Kitt Peak National Observatory, Tucson, Arizona, U.S.A.
Dr I. Elliott	Dunsink Observatory, Dublin 15, Ireland
Prof J.D. Fernie	David Dunlap Observatory, Toronto, Ontario, Canada
Dr J.R. Garcia	Instituto Copernico, Buenos Aires, Argentina
Mr R.M. Genet	Fairborn Observatory, Mesa, Arizona, U.S.A.
Prof D.S. Hall	Dyer Observatory, Nashville, Tennessee, U.S.A.
Dr P.W. Hill	University of St Andrews, St Andrews, Fife, Scotland
Dr I. Jankovics	Gothard Astrophysical Observatory, Szombathely, Hungary
Dr D. Kilkenny	South African Astronomical Observatory, Cape, South Africa
Prof E.F. Milone	University of Calgary, Alberta, Canada
Prof T. Oja	Kvistaberg Observatory, Bro, Sweden
Dr K. Olah	Konkoly Observatory, Budapest, Hungary
Mr T. O'Sullivan	Bord Telecom, Dublin 1, Ireland
Prof H.S. Park	Korean Astronomical Observatory, Taejon, Korea
Dr D. P. Pyper Smith	University of Nevada, Las Vegas, Nevada, U.S.A.
Dr F.R. Querci	Observatoire Midi-Pyrenees, Toulouse, France
Dr C.L. Sterken	Astrophysical Institute, Vrije Univ., Brussels, Belgium
Dr M.K. Tsvetkov	Bulgarian Academy of Sciences, Sofia, Bulgaria
Mr H. van Bellingen	Schull Planetarium, Cork, Ireland
Dr I. van Breda	Dunsink Observatory, Dublin 15, Ireland

One of the social events associated with the workshop was a walking tour of the mediaeval city of Kilkenny which started in front of the main gate of Kilkenny Castle, the former home of the Butler family, the Earls of Ormonde. Left to right are John Butler, Ed Budding, Mike Bode, Mary Crawford, Dave Crawford, Peter Baruch, Deny Baruch, John Baruch, Phil Hill, Sheila Hill, Russ Genet, Joyce Genet, Dave Kilkenny, Ian van Breda, Jamie Garcia, Maria Garcia, Huang Park, Tarmo Oja, Francois Querci, Diane Pyper Smith, Silvi Oja, Don Fernie, Istvan Jankovics, Yvonne Fernie, Eugene Milone, Katalin Olah, Helen Milone, Mimi Hall, Doug Hall and Ian Elliott.

Pictured here the following day in front of the 72-inch telescope constructed in 1845 by the third Earl of Rosse at Birr Castle are (left to right): Jamie and Maria Garcia, Alexander Antov, Ed Budding, Katalin Olah, John Baruch, Milcho Tsvetkov, Tarmo and Silvi Oja, Dorothy and Ian Elliott, Ian van Breda.

SUBJECT INDEX

accretion disc, 4, 101, 102, 103, 104, 142
accretion onto neutron star, 144
acquisition and guiding, 12, 13, 16-18, 22, 25, 42, 43, 139, 140
advanced X-ray astronomy facility (AXAF), 144
amateurs, vii, viii, 7, 142, 143
Apache Point telescope, 5
APT, ix, 6, 11-14, 15, 27, 60, 107, 139
Argentinian 40-cm telescope, 27-34, 139
Armagh Observatory, 95, 97
array processors, 132
Asiago Observatory, 7
asteroids, vii, 37, 72, 139, 141, 145
ASTRONET, 7
astroseismology, *see* stellar seismology
ATIS, 19, 43
atmospheric extinction, 3, 42, 45, 47, 48, 72, 85, 108, 109, 117-124
atmospheric seeing, 79
Automatic Monitoring Telescope, 7, 141

Baade—Wesselink method, 60
Belogradchik 60-cm telescope, 8, 69-74, 89, 92, 139
Berkeley automated supernova search, 144
Berkeley Automatic Imaging Telescope (BAIT), ix, 139
BGARNET, 89-93
BGNET, 92
binary stars
  eclipsing, viii
  interacting, 142-144
  orbital periods of, 142-144
  RS CVn, ix
  spotted, 50
  X-ray emitting, 143-144
Birr Castle 72-inch telescope, 145, 149
BITNET, 89, 91
black hole, 102, 103, 104, 105, 143
blazars, 101, 102, 103, 104, 105
Bradford robotic telescope, ix, 7, 21-26, 139, 140
brain, human, 133, 140
broad line region, 103, 105
buildings (*see also* individual telescopes and domes), 11, 16, 37-38, 78
BULPAC, 89, 90

C-11 twin telescope, 85
Capodimonte Observatory, 8
Carlsberg Automatic Meridian Circle (CAMC), viii, 4, 7

Carter Observatory, 15
Castelgrande Observatory, 8
cataclysmic variables, 72, 143, 145
catalogues (*see also* databases)
  GCVS, 65
  HST guide star, 63
  IRAS point source, 65
  SAO, 65
  STARCAT, 91
  Tycho, 65
Catania 80-cm telescope, 8
CCD
  advantages over pm tubes, 139
  cameras, 12-13, 22, 25, 36, 139
  control, 22, 23, 135
  data handling, 135, 137
  driftscanning, 126, 128, 130
  imaging, ix, 79, 126, 129, 139, 143
  photometry, ix, 72, 79, 113, 139, 144
  spectroscopy, 139
Celestron "Compustar", 15-20
Chappuis bands, 18
circumstellar matter, 142
colour indices, 108, 118, 121
comets, 36, 37, 139. 141
Compton scattering, 102
computer interface, 36-37, 70-72
computers, 13, 22-23, 36-37, 70, 78, 89-90, 128-138
condition monitoring, 5, 38, 136-137
control systems, 22, 23-26, 43, 70, 78-79, 131-138, 140
corner stars, 63-68, 145
COSMOS, 145
cost considerations, 78-80, 81, 82-84, 125
cryogens, 123, 141

Danish 20-inch telescope, ix, 7
databases (*see also* catalogues)
  distributed, 136-138
  INCA, 65
  SIMBAD, 65-66, 91
DATAEXP, 89
dewing, 16, 19, 22, 37
differential rotation, 50-51
distance scale, 142, 143, 144
domes (*see also* buildings), 11, 70
drives, 5, 16, 23, 25, 36, 42-43, 45, 47, 64, 70

Subject index

dust
  formation of, 59, 103
  circumstellar, 142
  emission from, 59, 102, 103, 104, 142, 143, 144
dwarf novae, *see* cataclysmic variables
Dyer Observatory, 43

electronics, 13, 36, 131, 133
El Leoncito Observatory, 31
environmental monitors (*see also* weather stations), 6, 22
E regions, 113
EUNET, 89
EURONET, 92
EXOSAT, vii

Fairborn 10-inch telescope, 42
Fairborn Observatory, 43
Fastnet telescope, ix, 35-38
field correctors, 125-126, 129
field of view, 35, 36-37, 79, 125-126, 129
filter sets, 108, 109
flashes (in sky), 128
flickering, 72
Forbes effect, 118, 120
French Revolution, 109
Future Small Telescope, 140

galaxies, 37, 72, 102, 111, 125, 128, 134, 143
  active nuclei of, x, 101-106, 144
  elliptical, 108
  IRAS, 102, 103
  M31, 143
  mergers of (*see also* galaxies, starburst), 108
  radio, 102
  Seyfert, 102
  spiral, 103
  spiral arms, 108
  starburst, 101, 102, 103, 104, 105
GEONET, 92
Global Network of Automated Telescopes (GNAT), 33, 77-84, 85, 145
gravitational lens, 144

HII galaxies, 102, 103
HII regions, 102
historical development, viii, ix
Hubble constant, 144

Hubble—Sandage variables, 144
Hubble Space Telescope, 126-128, 144

image processing, 22
infrared detectors, 123
infrared extinction, 117-124
infrared observations, 102
infrared photometry, *see* photometry, infrared
Infrared Space Observatory (ISO), 144
International Ultraviolet Explorer (IUE), 95, 97
INTERNET, 92, 97
interstellar medium, 143
interstellar reddening, 108, 109

jets, 102, 103
JET-X, 8
Jodrell Bank, 96
Johnson passbands, 25, 122, 123

Kitt Peak National Observatory 50-inch telescope, viii
Kitt Peak National Observatory No. 4 16-inch telescope, 42-43
Konkoly 1-m telescope, 8
Kotipu Place Observatory, ix, 15-20, 139
Kurucz model atmospheres, 119

La Palma, 6, 7, 96
Leuschner Observatory 20-inch telescope, 144
LIDAR, 128
light echo method, *see* reverberation mapping
line blanketing, 103
LINERS, 102
Liverpool John Moores University, 140
local area network (LAN), 92

magnetically active variables (*see also* starspots), 142
Mauna Kea, 120, 121
mesures usuelles, 110, 111
METEOSAT, 87
metric system, 110, 111, 140
metrological reform (of photometer systems), 107, 112
mirrors, 11, 12, 27, 35, 78
  automated cleaning of, 121, 123, 141
  liquid, 125-130, 140
  tests of, 126-128
MODTRAN, 119, 120, 123
monitoring programmes (AGN), 104, 105
multi-frequency campaigns, vii-viii, 95-97, 101, 108, 144
multi-site campaigns, 77-84, 85-88, 95-97

Subject index

networks, x, 73
  Bulgarian computer (BGARNET), x, 8, 89-92
  neural, 133
  telescope, 33, 77-84, 85-88, 107, 112, 141
neuron chip, 134
New Generation Small Telescopes (NGSTs), 77-84, 140
noise
  photon, 41, 47-48
  scintillation, 41, 47-48
non-thermal emission, 102, 104, 144
novae, vii, 142, 143

observing constraints, 95-97
optical fibres, 13, 129, 140
optically violent variables (OVVs), 102
optics, *see* mirrors
outer planets, inner satellites of, 141

Palomar Observatory Sky Survey, 63, 68
parallel processing, 131-138, 140
patrol work, 37, 38, 71-72, 144-145
pattern recognition (*see also* acquisition and guiding), 5
Perugia 40-cm telescope, 8
Phoenix 10-inch telescope, 15
photometric standards, 115-117, 140
photometric systems (metrological reform of), 107, 112
photometry (*see also* CCD)
  colour equations in, 113-114
  data reduction in, 71
  differential, 32, 33, 65-66, 71, 85-88, 108, 139
  high-precision, 41-48
  high-speed, 32, 33
  infrared, 117-124, 140
  intermediate band, 108, 111, 142
  millimagnitude precision, x, 7, 33, 86, 107, 108, 118, 140
  multichannel, 7, 85
  photoelectric, 4, 11, 17, 41-48, 69, 72
  photographic, 108, 110-112
  transformation problems in, 118, 111-112
  visual, 108, 110-111
  wide band, 79, 108-109, 111, 115-117
  wide field, 36-37
pipelining, 131-132
planetary nebula central stars, 86
planetology, 85, 143-144
plate archive, 92
pointing-flow, 24-25

polar axis, alignment, 16
pre-main-sequence stars (*see also* young stellar objects and stars, T Tauri)
pulsars, 145

quasars, vii, 102, 105, 111, 125, 128

RADAUS, 89
radio astronomy, 79, 101
radio galaxies, 102
radio telescopes, 4-5, 96
Rayleigh scattering, 118
reverberation mapping, 103
ROSAT, 96, 144
Royal Greenwich Observatory, 115, 140
Rozhen Observatory, 69, 72, 89

St Andrews twin photometric telescope, 63
Sanbothay Gothard Observatory, 8
scaling law (of telescope costs), 82-84
scheduling, 25, 78
Schmidt camera, 35-37
science education, 79-82, 144-145
SCOPES, 95-97
Sierra Nevada Observatory, 6
sites *(see also individual telescopes)*, 79-80, 87
Skillman 12-inch telescope, viii
software, 13, 18, 38, 95-97, 132-133
solar system bodies (*see also* asteroids, comets), 18, 111, 141
South African Astronomical Observatory, 11-14, 96
Spaceguard, 145
space satellites, 5, 80, 82, 85, 86, 87, 95
spectrograph, 135
spectrophotometry, 87, 105
spectroscopy, 126, 129, 139, 140
STARLINK, 5, 23, 96
stars (*see also* binary stars, novae, supernova, white dwarfs)
  algolide, 85
  Am, 86
  Ap, 86
  Be, 86, 143
  blue horizontal branch, 145
  carbon-rich, 59-61
  Cepheid variable, 142
  chromospherically active, 41
  clusters of, 36-37
  flare, 18, 72, 73, 74, 96, 128

stars - *continued*
  formation, 115
  helium, 142
  hot subdwarfs, 145
  OB supergiant, 143
  pulsational variable, 59-61
  pulsating degenerate, 145
  RCB, 59-61, 142
  red giant, 72, 85, 86, 143
  roAp, 34-36
  RS CVn, *see* binary stars
  RV Tauri, 142
  supergiants, 59
  symbiotic, 72, 107, 108, 143-144
  T Tauri, 86
  UU Her, 59-61, 142
  W CMa, 86
  Wolf—Rayet, 107, 144
  δ Scuti, 19, 32-33, 85
starspots, x, 41, 142
  lifetimes of, 49-58
  migration curve for, 51
star tracker, 4
stellar occultation, 86
stellar pulsation, 142
stellar rotation, 49-51, 142
stellar seismology, 4, 6, 32, 86, 144
stellar winds, 102, 103
sunspots, 55-56
supernovae, vii-viii, 102, 103, 105, 111, 139, 144-145
supernova remnants, 102, 105
Sutherland Automatic Telescope, 11-14, 140

targets of opportunity, vii-viii
telescope, mounts, 27-31, 35, 78
  background emissions from, 123
  drives for, 30-31
  transit, 126
telescopes, new generation, 77-84
telluric atmosphere, 86
Tenerife Observatory (Izana), 6
thermal emission processes, 102, 104
time critical events, 80-81
time-line, 96
time synchronisation system, 70
tracking (*see also* acquisition and guiding), 126, 133
transformation coefficient, 45-48
transputers, 13, 134-136

UK Schmidt Telescope, 63, 145
ultraviolet observations (*see also* International Ultraviolet Explorer), 95, 104-105
undergraduate students, 7, 80-81, 144-145
UNICOM, 90
users, number of per telescope, 83

Vanderbilt—Tennessee State 16-inch telescope, ix, 41-48
Virgo cluster, 144
VLT, 6, 8
von Neumann computer architecture, 132
von Neumann segments, 132

Wainscoat—Cowie filter, 123
warmers, 102
water vapour, absorption of IR radiation by, 117, 119, 120, 121, 122
weather station (*see also* environmental monitors), 5, 19, 22
white dwarfs
  in classical novae, 143
  oscillations, x, 145
Whole Earth Telescope, x, 143
William Herschel Telescope, 135, 140
Wisconsin 8-inch telescope, vii

X-ray observations (*see also* binary stars), 95, 101, 102, 103

young stellar objects (*see also* T Tauri stars, pre main-sequence stars), 102, 104

Zenith tube, vii, 3

3T1M project, 85-88, 139, 141